Doctors and Healers

———————

Doctors and Healers

Tobie Nathan
and Isabelle Stengers

Translated by Stephen Muecke

polity

First published in French as *Médecins et sorciers* © Éditions La Découverte, Paris, France, 2012

This English edition © Polity Press, 2018

Polity Press
65 Bridge Street
Cambridge CB2 1UR, UK

Polity Press
101 Station Landing
Suite 300
Medford, MA 02155, USA

ISBN-13: 978-1-5095-2185-2
ISBN-13: 978-1-5095-2186-9 (pb)

A catalogue record for this book is available from the British Library.

Library of Congress Cataloging-in-Publication Data

Names: Nathan, Tobie, author. | Stengers, Isabelle, author.
Title: Doctors and healers / Tobie Nathan, Isabelle Stengers.
Other titles: Medecins et sorciers. English
Description: Medford, MA : Polity Press, [2018] | Originally published in
    French as: Medecins et sorciers. | Includes bibliographical references
    and index.
Identifiers: LCCN 2017057449 (print) | LCCN 2018005531 (ebook) | ISBN
    9781509521890 (Epub) | ISBN 9781509521852 (hardback) | ISBN 9781509521869
    (paperback)
Subjects: LCSH: Medicine--Philosophy. | Quacks and quackery.
Classification: LCC R723 (ebook) | LCC R723 .N33713 2018 (print) | DDC
    610.1--dc23
LC record available at https://lccn.loc.gov/2017057449

Typeset in 11 on 14 pt Sabon by
Servis Filmsetting Ltd, Stockport, Cheshire
Printed and bound in Great Britain by Clays Ltd, Elcograf S.p.A.

For further information on Polity, visit our website:
politybooks.com

Contents

v

Editor's Note

This is a translation from French of the revised
2012 edition which saw the addition of the final
two chapters, 'Users: Lobbies or Political Creativity?'
and 'Doctors, Healers, Therapists, the Sick, Patients,
Subjects, Users . . .'. These new chapters are based
on a presentation to a symposium, 'User Responses
to Psychotherapy', organized by the Centre Georges
Devereux, Paris, 12–13 October 2006.

1
Towards a Scientific Psychopathology

Tobie Nathan

I   THE BENEFITS OF FOLK THERAPY

Today we continue our investigations with Zézé about the origin of the masks, but this time he appears not to understand.

'The spirits of the bush must have existed at the very beginning of the world?' I ask him.

Zézé looks at me as if puzzled by my stupidity.

'No, no, of course not,' he says, shrugging his shoulders. 'We, the Zogui, did all that.'

He explains the meaning of this word, with which I had not been familiar. 'Zogu' literally means 'man', but in this particular sense 'the great fetisher, the master of the spirits of the forest, the complete man'.

'In the beginning of the world,' he adds, 'there was only water, the serpent and two medicines: Belimassai and Zazi.' These two words in fact mean the same thing: lightning stone. But the first is used only by men and the second by women.

Pierre-Dominique Gaisseau, *The Sacred Forest: The Fetishist and Magic Rites of the Toma*, trans. Alan Ross, London: Weidenfeld and Nicolson, 1954, p. 99 (translation modified)

Scientific therapy and folk therapy

Psychotherapy that is called 'scientific' (obviously, I'm not talking about its truth value, only its method, whether Freudian, anti-Freudian or neo-Freudian, fanatical Kleinian or crypto-Lacanian, whatever . . .) – this type of psychotherapy, I was saying, always contains a single premise that is clear and explicit: humans are alone! We are alone in the universe, therefore alone in the face of Science, and consequently also alone in the face of the state. It is by way of this unique formula that I am able to summarize the articles of faith of the 'science'-based psychotherapies that I know. Since the second half of the nineteenth century, all the theoretical systems that have seen the light of day flow from this idea.

And I do mean all the theoretical systems, because once you take a close look, even from a logical point of view, there is only one class of psychotherapies since they all derive from the same basic premise:

1. Madness is a kind of 'illness'.
2. Like all illnesses, it resides in the 'subject': in its psyche (psychoanalysis and its countless offshoots); in its biology (psychopharmacology); in the layers of its unique history ('existentialism'); and in its educational repercussions ('bioenergetics', 'gestalt therapy', 'transactional analysis').

Now, let's imagine an astute critic – let's say a literary type, a humanist – no doubt interested in psychotherapy, but also in cultural differences, religions and their histories, in philosophy. Let's put this person into dialogue with me. I am sure he or she will immediately say: 'I have no idea what you are talking about. Are we not all alone, just as we are, alone in relation to ourselves, to our destiny, to our death . . . so what kind of solitude are you talking about? Explain yourself!'

The Benefits of Folk Therapy

Solitude

A woman faints! Think first about the solution that societies like ours offer, societies with one universe. We have to think that this woman is suffering a disorder that is part of the knowable world; let's call it hysteria for the sake of simplicity. By starting to think like this, we think of her imbued with sexual drives that she (she alone?[1]) doesn't recognize. So, experts are called in – I'm not so naive that I think them particularly erudite, I simply define their function, mastering rational knowledge. So a scientist will unmask her unconscious strategies, lead her to become conscious of them and help her work out a more mature strategy for existence. Let us think for a moment. The gaze we bring to bear on this woman simplifies her character ('regressed', as they say). One can feel sorry for her ('she is in pain'); lecture her ('she is infantile'); help her out of Christian feeling (Christian . . . Christian? Do we need to be reminded that there are sick people who are a bit suspect, *pas très catholique*, as we say?); guide her, morally, out of 'humanity', out of duty in any case (God only knows what is set loose in sick people by these kinds of interests). Whatever the case, she remains alone – alone in the face of the 'scientist', therefore in the face of medicine, and of the state.

In order to handle one and the same fact, namely a woman fainting, societies with multiple universes never fail to propose a solution that consists of the postulate that she has been taken over by a spirit. On the basis of this fact, and quite logically, it becomes indispensable to call for the help of a 'connoisseur of spirits' – *master of the secret,*[2] *master of the knowledge*

	One universe societies	Multiple universe societies
Cause	'Sickness' – hysteria	'Spirit attack'
Aetiology	→ Unconscious sexual drives	→ Intentionality of the supernatural being
Social Actor	Master of rational knowledge	Master of hidden knowledge
Treatment philosophy	• Unmask unconscious strategies; • Work towards awareness, towards a maturing of under-developed capacities	• Identify the invisible being • Recognize his intentionality • Negotiate with him • Set up permanent 'shopfronts' to trade with his world
Representation of women	• Obsessed by sex • Infantile • Regressive	• Unconscious informant of an invisible world that is good to know • 'Witch' • Chosen one • Interpreter
Therapist's attitude toward women	→ 'Humanitarian', cares for women • Feels sorry for her, lectures her, supports her, helps her…	→ Doesn't care about the woman, but the spirit • He looks away, goes off to 'interrogate' the hidden…
Results of the treatment	→ Loss of reference group • Falling back on medicine and the clergy • Isolation	• affiliation to a new groups • assignation to social sites

Figure 1.1   A woman faints

to be gained through initiation. If he officiates according to his art, this woman will necessarily become the unknowing informant of an invisible world, one that is worth knowing about. She is an ambiguous character, potentially having multiple personalities. We can scoff at her (strange kind of person, wouldn't you say?), fear her (she's also a kind of 'witch'), envy her (she has been chosen), and interrogate her (she is the interpreter of the hidden, a woman who straddles two worlds). As soon as disorder erupts, it is useful for the whole group; it helps them complicate the world and learn about what remains invisible to it.

The deliberate isolation of patients runs through the whole of psychological and psychopathological thought; it's even one of its implicit presuppositions. Consider another example for a moment: a child who does not speak, rocks continually and makes strange and incomprehensible noises. As you know, these children like playing with water and sand, don't sleep at night and have a curious preference for the company of mature men – perhaps even grandfathers – rather than that of their mothers and women in general. What would our scientist say? He or she will always come to the conclusion that the poor child is lacking, that he or she has stumbled on his way up through the stages that each child has to climb before raising themselves to the human level. He has stopped at 'symbiosis', they will think, or at 'primal orality', or more generally at pre-genitality ... But the master of secret knowledge will always come to the conclusion that the child has a 'singular nature'. This child is silent with other humans, but the 'master of the secret' will claim that this child has silent interlocutors and a specific essence. Douala, one of my patients who came from the Cameroons, was said to be a 'hippopotamus child'.[3] Consequently, he was deposited on an island in the middle of the river so that his fellow creatures could come to pick him up as one of their own. Perhaps the hippopotamuses didn't want to deprive my patient's parents of such a pretty baby. The Douala family no doubt reasoned that the hippos took their own and gave back the human child . . . The fact remains that, once the child had been recovered from the river, he slowly overcame his mutism . . . .

'You are being selective about the facts you use,' my well-informed critic will retort. 'No doubt you are led astray because of your biased view of African cultures. What's all this about multiple universe societies? Can't we imagine there must be something there which our science can patiently explore? Can't this thing, that we name a "hysterical" illness just for the sake of temporary convenience, be considered a possible avatar for all human development? Isn't it the same thing that certain scientifically underdeveloped peoples still interpret metaphorically as an attack by imaginary beings? We have successfully drawn up nosographic tables, given them shape and checked their consistency – so aren't we much better off? Isn't it better to think of the child you are speaking of as "autistic" and humbly accept the limits of our therapeutic powers? As far as I'm concerned, I think there is a kind of moral grandeur in recognizing the limits of our omnipotence.'

'You are so damn naive! Have you thought for a moment of the fact that a statement of that type – I agree, the most common in our profession – implies that you think there are irrational peoples in the world, with "pre-logical minds",[4] awash in a maelstrom of emotions, incapable of conceptualization, following only their natural impulses? Every day, I deal with people from such cultures; I meet them regularly and I can assure you that reason is distributed in the same

	One universe societies	Multiple universe societies
Cause:	Sickness	Specific nature of the child
Aetiology:	*relating to lack:* • Biological • Interactive (lack of mother-child interaction) • Libidinal fixation	*relating to surplus:* • Ancestor child • Witch-child • 'Hippopotamus child'
Social Actor:	Master of rational knowledge	Master of hidden knowledge
Intervention Philosophy	• Attempt to 'repair the lack' • Philosophy of affective re-education • Isolation of the sick child among so-called similar cases	• Identify the nature of the child • Study the kind of language current in the special universe that his singularity reveals • Establish permanent 'shop fronts' to trade with this world

Figure 1.2   A child not speaking

way over there as it is here. I even feel a little stupid reminding you of this as it is so obvious! Honestly! We have no choice but to think that aetiologies – whether of "primitive" or scientific origin – are, as I've said, *all* rational. They can only be distinguished by the fact that each triggers a different action on the world. This is why I think that the so-called "scientific discoveries" of Professors Charcot and Freud, condemning witches, sibyls and pythonesses to the misery of hysteria, are nothing but the official stamp on the death certificate marking the demise of multiple Universes – a statement of failure, in some ways . . .

Diagnostics or divination

In any case (has it not been said often enough?), from the very beginning of any therapeutic activity, the "master of the secret" invades the world. He doesn't interrogate the "sick" person, just the objects related to the hidden universe. He asks the sand, shells, a palm-nut rosary[5] and the Koran. Sometimes it's enough for him to "see", thanks to a "gift".'

'Come on!', my critic will no doubt interrupt me immediately. 'You don't actually believe in tarot cards and other such clairvoyant stuff?'

'Not so fast. Why do you already want to cast aspersions? Wait! Let me elaborate on my idea . . . If one submits to the kind of investigation that you seem contemptuous of, then disorder is necessarily seen in a particular way. *It then becomes a sign* of an obligation to be interested in the richness of the world and the multiplicity of beings that inhabit it. In these worlds, disorder always ends up being a tangle in the lines of communication, a crossroad, just at the point where the universes are superimposed . . . Ah, my friend, you have to eliminate the words "belief" or "believe" from your vocabulary. Take my word for it, no one, anywhere, believes in anything! A divinatory apparatus is always *a creative act*. It institutes the interface among universes; it makes them palpable and then thinkable. So, will you carry on as you were, telling yourself that these systems are made out of naive thinking, founded on the "childish" credulity of ignorant peoples? For my part, I'd rather think of them as unleashing an extraordinarily complex machinery designed to create links, a consummate art for multiplying universes. Because such inquiries, basically directed towards what is hidden, displace any interest centred on the ill person (as always, prone to stigmatization). They displace him or her:

1. towards the "invisible";
2. from the individual to the collective;
3. from the inevitable to the reparable.

But for that to happen we still need the existence of a hidden world, a secret world, known only to the *masters of the secret.*

On the other hand, the scientist, as you know, investigates symptoms, naturally via the intermediary of the patients themselves, because no illness can escape the one real world, that described by academic psychopathology. I have recently discovered that scientific research is never trying to discover worlds, just to extend its own. In our universe, if it occasionally happens that we think that some disorder is not known, we still deem it to be potentially knowable. Perhaps the scientist will discover it one day and give it his name – "Charcot's Disease", "the Bleuler Syndrome".

It is for this reason that all cultural worlds with multiple universes have recourse to divination while all those with one universe use diagnostics.'

'That's an interesting suggestion; it gives me food for thought. Can you tell me more?'

Statistical categories vs real cultural groups

'I could add that, when he starts divining, the master of hidden knowledge has an implicit aim. This is to find out about the sick person's unexpected membership attributes and thus ultimately to assign him or her to a group. For example, a particular child might be someone exceptional, likely to "eat" his own parents.[6] No one knew it at the time, but when he was in his mother's womb he was accompanied by a twin whom he devoured when they were foetuses. So he belongs to the large inter-ethnic[7] family of twins, those obscure beings that are best protected, respected and honoured if one is not to ask for trouble. Rest assured, this sufferer will emerge from his course of therapy having discovered a new sense of belonging, thinking of himself in the fellowship of twins. He will submit to the protective rituals of his new group and will respect the special dietary restrictions, etc.

On the other hand, the aim of the "scientist" is always to cut the subject off from his universe and his possible affiliations and also to submit him, just like everybody else, and especially as a sole individual, to the implacable and blind "laws of Nature".[8] But what is the scientist working on here? What objective does he have in mind as he suppresses all real groups – twins, those possessed by Yoruba divinities like Ogún, Shangó or Sakpatá, sorcerers, witch-hunters, ancestors – all these groups that constitute indispensable links in the elaboration of therapies? The answer seems obvious to me: it is simply a case of him increasing his clientele. Because when it comes to psychopathology, medicine and its derivations, wherever it sets itself up with its foot soldiers (doctors), its quartermasters (pharmaceutical laboratories), its judges (the scientists who sort the "real" from the "false", what exists and what is in the "imagination"), it always has the effect of breaking down memberships. As soon as you set up a dispensary in Bamako, you will no longer see any Bambaras, no more Dogons, nor any Puels; just "subjects", who quickly become empty envelopes; they are subjected. They are "hooked" on prescriptions; they are Largactyl-Nozinan-Anafranil-Prozac "junkies".'

'Ah, you are annoying me with your military and third world talk,' my critic replies. 'Don't you think that you are spoiling an argument that began nicely enough by corrupting it with these criticisms of the medical establishment that must look a bit suspicious?'

'Not at all. But I don't like the kind of soft thinking that provides nothing to react against. I am simply trying to set out my ideas clearly. It is no doubt for this reason that they sound polemical to you. So, consider a piece of factual evidence: the psychopathological categories that are at the basis of how psychiatrists (and I include psychoanalysts and psychotherapists here) classify their patients never have their origins in real groups. Have you ever heard of groups of "obsessives" or "paranoiacs"? Have you spotted them in the same place, getting ready to undergo the same ritual therapy, recognizing each other as members, and perhaps – who knows? – even with a common ancestor? Do you know of a *temple* for the "hysteria" entity, or an *altar* for essence of "schizophrenic"? Of course not, since psychopathological categories are disjunctive concepts that only bring individuals together for statistical purposes. I can tell you, since I have frequented quite a few psychiatric institutions, that I have always heard patients complaining about being mixed with *mad people*. ("What am I doing here? Everyone is mad . . ."). And the doctors make fun of their carrying on, cleverly recognizing supposed "denials" of the illness. . . . In fact, and one always has to listen carefully to the exact words uttered, people admitted into psychiatric wards explicitly complain about not recognizing the group in which they are statistically classed.

This is why (believe me, I have had the experience) you can legislate all you please on the freedom for the patient to see their case notes. In psychopathology, the diagnoses, i.e. methods of extracting people ("subjects") from their group, always remain secret.'

The construction of *Truth*

My astute critic has the usual rejoinder: 'Give me a straight answer! You know very well, as an "institutionalized" intellectual, that the "spirits" called upon by the healers don't exist. Or at least you don't believe in them yourself.'

'My dear fellow, I'm afraid I have to answer that I can't allow you this criticism. And I have at least two arguments to make the case:

1. First, allow me to tell you that I am hearing you speak like a divinity, not like a human! You seem to develop a thought without premises. It is very strange to ask, "Do spirits exist?"

    In a single-universe world, the existence of spirits is clearly ridiculous. Just imagine spirits having problems putting on their shoes, taking a bus or waiting in a queue to order a hamburger. This would certainly be funny, but it's absurd. Spirits have irreducible qualities; they can only be evoked in a world of multiple universes *since their very evocation in and of itself calls the second universe into existence*. And here is my second argument:

2. So you research diagnostics about nature, you report on existence, you find proofs. I am a relationship technician and, like any practitioner, I care most about effectiveness. From this position, I have learnt about the extraordinary release of creative – hence life-producing – energies brought about through multiple universe systems. A patient is like a stone. At first sight he seems monolithic, whole, perfectly smooth. Hasn't he learnt to work at will on any imperfections? Launch into interrogation on the hidden and you will see him crack along his own specific fault lines right in front of you. *If it is necessary to appeal to spirits to trigger this system, then spirits certainly exist, at least in as much as they are the invisible heart of the setting.*'

'So be it', my enlightened critic will now say. 'I am sympathetic with the second argument. Although I'm not a professional, I can imagine the suffering of someone whose job is healing, and I really want to concede that a professional prefers a system that works, even if it isn't approved, to some other system that he knows is dysfunctional, even if this latter has been blessed by universities and churches. We know of such things in the past. So what kind of psychopathology are you advocating? Aren't you really making a case for a return to the past, to charlatans, bonesetters, street performers, acrobats?'

Risky psychopathology

'My dear fellow! You must be aware of the ideas of our invaluable Isabelle Stengers on the characteristics of a science. You remember that she shows that a science is the activity of a group of scientists who have agreed to take a risk – I should say, agreed to submit their thought to risk.[9] Answer this question honestly: to what risks are our psychopathologists exposing their thought models? Now tell me, who has any device to hand that could possibly contradict them? They decree the existence of an object that only they are able to perceive; they alone make the instruments designed to describe the object and make it opaque to any outsider; then they themselves validate the adequacy of their instrument. So the loop is closed, and even padlocked. Here we have thought without any risk. But the master of secret knowledge, using divination rather than diagnostics, exposes himself permanently to risk, and first of all by being contradicted by the real expert which the sufferer de facto becomes in this setting. Try carrying out such a curious experiment yourself by making a clear and unambiguous statement about someone. Say to your tobacconist, "Sir, I 'saw' that you are the oldest of five brothers . . ." Go on, do it! Do it and you will feel your stomach tighten when he says to you, "No way! I am an only child!" I imagine your feelings being far more perturbed if he replied, "Oh! But how do you know?" It's only through such an experiment that you will understand, and from the inside, the way divination puts your thinking at risk.

	One universe societies	Multiple universe societies
Method	Diagnostic	Divination
Social Actor	Expert	Chosen one
Place of investigation	In the patient	In the deviner
The patient is...	a bearer of 'illnesses' and of 'structures'	an expert
Philosophy of the method	• Detailed investigation of the 'visible', of the perceptible, of the measurable • Extension of the one universe	• Displacement of interest: – from the visible to the invisible – from the individual to the collective – from the fatal to the reparable
Results of the treatment	• Assignment of the subject to statistical categories • Isolation in the middle of so-called peers	• Creation of interfaces among universes • Announcements about new memberships • Affiliations to secret groups

Figure 1.3   The construction of truth

Now describe to me what intellectual risk a psychologist crazy about the Rorschach test is taking, or a psychiatrist obsessed with the *Diagnostic and Statistical Manual of Mental Disorders (DSM)*. Recourse to these instruments has the sole aim of disqualifying other types of experts: the ill person, their family or their environment.

In my opinion, psychopathologists need to take risks to bring about the creativity that is an indispensable characteristic for building up a scientific account of things. What happens in societies with multiple universes is very telling on this point. For example, if I start to divine, the locus where the drama of knowledge plays out is myself. People ask, "How does he know? What method is he using?" But when I do a diagnostic procedure, the drama is happening within the patient in the absence of any witness likely to be called upon to testify. That's just one simple example. Obviously, I don't expect that simply replacing diagnostic methods with divination strategies would allow psychopathology to rise to the status of a science (though it would help a lot). I just want to draw attention to the fact that these "wild" practitioners would be more inclined to get involved in "risky" procedures.

In short, any psychopathology that is interested in the sick, whose main concern is to objectify "illnesses", necessarily distances itself from the tensions that permit a science to be constructed. So, to answer your question, I do advocate a psychopathology that takes risks, that makes a really fine-tuned description of therapists and therapeutic techniques, but not of the sick people. Because in this domain, all that can be observed are the therapists and their objects – and of course I mean all the objects: tools, but also theories, thoughts and even supernatural beings . . . .'

'Speaking of which, tell me something: these supernatural beings – forgive the loose expression – are specific for every human group. I believe some of them are located in the water of rivers, others deep in the dark forest, yet others in underground tunnels or in abandoned dwellings . . . So you would have to be au fait with each group and with each specific modality of interaction with these invisible beings. And in addition, you would have to work in the patients' mother tongues because I don't think that the names and characteristics of the spirits would translate easily. Would a psychopathologist have to know many languages, cultures and specific modalities to enter into relations with spirits? Come on, be reasonable! Your position might be intellectually seductive, but it is totally unrealistic, I'm sorry to say.'

'I'm surprised how willing you are to oversimplify. Any naive person can easily understand that when it is a matter of modifying someone's whole being, this is only possible from within their language, along with its referents and its divinities.[10] There lies, I think, both the greatness and all the difficulty of our profession. Before establishing "general laws" on the nature of disorders, psychopathology should first get busy for each culture, describing systematically the activities of a certain category of person that their group has entrusted with modifying the internal functioning of other people. We condescendingly call these people "healers", reserving for ourselves the noble term "doctor". But:

1. they are in fact our colleagues;
2. they are the repositories of knowledge that we first have to acquire before we can aspire to be scientific at all.'

'Ah, now I know what you are up to! You have a way of turning things around . . . . So, according to you, the healers retain real knowledge, while psychopathologists are struggling with ideological thought? Isn't that what you are thinking?'

'Naturally, since we are talking about "technical knowledge"! Who would have it, if not healers? They are virtuosos who have refined their know-how over millennia.'

'Could you tell me more about technique, then? I challenge you to tell me how you go about making these systems function that are so foreign to your training, and especially in the middle of an academic context.'

A clinical illustration

University Paris-VIII, the Centre Georges Devereux. One Tuesday morning at 11. A large hall with a very high ceiling . . .

'Imagine the scene . . . My friend Lucien was there, my Yoruba brother, whose people – at least that is how I like to think of it – left Egypt a few millennia before me. He is a psychologist and a psychoanalyst with French university qualifications. Yet no one is better than him at handling the subtleties of the Yoruba, Fon, Goun, Adja, Mina and Ewe languages.[11] At one time or another, we have all been impressed when, during a session, he had occasion to plumb the depths to come up with formulae that, according to the Yoruba expression, his grandfather "had once instilled in him", and then his eyes become as red as burning coals. Then there is Hamid, a psychologist from the Kabyle. He is spiritual, thoughtful, deep and passionate, most of all about the subtle nuances in the Kabyle, Berber and Arabic languages. There is also Marième, a Peul from Senegal who can't listen to a patient without translating the words, deep in her heart, into Wolof.[12] There is Alhassane, a Peul from Guinea, so thin he looks like a shadow projected onto the ground. He can follow Malinké, Soussou, Bambara, Manding, Khassonkhe and Peul. And Geneviève, a vivacious Lari from the Congo. I am certain she would give years of her life not to miss out on the appearance of a forest spirit. She speaks Kikongo, Lari, Lingala, a little Kiswahili,[13] and Sango.[14] Today she is probably thinking, as on every occasion that we talk about exile, of earlier days in Brazzaville, of the croaking of frogs at night after a storm when one savours the richness of a complex and multiple world.'

'Fine, fine . . .,' my critic interrupts. 'Let's get to the point. Just how many of you were there consulting with this patient?'

'I can't remember! At least a dozen, all with French qualifications. Psychologists and psychoanalysts, of course, but also doctors, anthropologists . . . . Next to us was nineteen-year-old Bintou, a magnificent Malian young woman of Bambara ethnicity. She was dressed like a fashionable young French woman, jeans and polo shirt – stunning! She told us about the long-standing issues which she kept bringing up with doctors and welfare agencies. She felt she was going blind, fainted for no reason, wandered like a lost soul from her aunt's house to her older sister's, from a youth centre to a troublesome grimy squat . . . . But the strangest thing was that Bintou had been pregnant at fourteen and hidden her condition from all and sundry. She gave birth alone in a toilet and placed the newborn on a second-floor windowsill. The infant fell and miracu- lously survived but was handicapped for life – blind, deaf and mute – and had been held in a specialized institution ever since, and no doubt autistic as well by now . . . Bintou was seated next to me. She dared not look at the group. I introduced everyone, specifying their cultural origin and their qualifications. We took our time . . . we could have been under a sycamore in a village on an afternoon, starting to chat . . . Marième, who had already had occasion to meet Bintou one on one, remarked, "We are trying to sort things so that a relative can take care of her because all she does is go from one family to another. This wandering has to be stopped." And then her case-worker explained that Bintou had originally been charged with attempted infanticide, then designated by the judge as a victim of child molestation.[15]

So, here we are in a real-life situation. We could have thought of Bintou as battling against "self-destructive fantasies", or "consumed with guilt", forged over time by her "borderline personality" . . . maybe she could undergo psychotherapy or psychoanalysis . . . But can psychoanalysis fix the irreparable damage caused to the infant? Can it explain why this infant survived, shadowing her mother with endless silent accusations? In such a situation, I insist, the therapist takes on a huge responsibility. If you decide to look for a disorder *inside* Bintou, you are bound to condemn her to keep up her solitary wandering in a world with a single universe. But if you start in a divining way, looking for invisible beings – and they must be Bambara invisible beings – at the origin of the disorder, this will necessarily pull Bintou out of her solitude.'

'Tell me now, instead of always philosophizing, how are you going to go about getting Bintou out of the bind she is in, caught between accusations of infanticide and the attentions paid to her by the welfare people?'

'I was the one who first started enunciating what sounded like statements of truth. I started by saying:

> *Tobie Nathan*: When Bintou was a little girl, she played with the big kids. She even looked adults in the eye. (*To Bintou*) Isn't this the reason that your mummy sent you here, because you were "special"?[16]
>
> *Bintou*: I don't know. One day I heard her speaking to someone about my departure, that's all . . .

And I added, now speaking to the whole group:

> *Tobie Nathan*: When she was one, and her father died, Bintou fell ill. (*Turning to Bintou*) Did you fall sick at that moment?
>
> *Bintou*: I don't know, I was one.
>
> *Tobie Nathan*: Your mother didn't tell you.
>
> *Bintou*: No. She doesn't tell me anything like that. She just sends me "things to wash with".[17]
>
> *Tobie Nathan*: I even know that you have marks on your body, traces each side of your stomach, I am sure of it . . .
>
> *Bintou (her eyes now bright with surprise, a smile on her lips)*: Yes, I have marks. There.[18] (*She shows the sides of her stomach*).
>
> *Tobie Nathan (I address the group)*: When her father died, she was sick for at least three months . . .
>
> *Hamid*: She even almost died.
>
> *Tobie Nathan (to Bintou)*: You take after your father, don't you, at least physically?
>
> *Bintou*: I don't know.
>
> *Tobie Nathan*: Really? Don't you have photos of your father?

Bintou: My mother doesn't want me to look at them. . .

Tobie Nathan (to the group): Bintou thinks a lot about her father.

Bintou: Yes, I think a lot about my father . . . every day . . . and it frightens me . . .

Tobie Nathan: And yet, when you were little, you often injured yourself. You came home with bloody knees.

Bintou: Yes, my mother says that I tired her out.

Tobie Nathan: What do you think, Lucien?

Lucien: That's the gist of it. You highlighted the important things, I think . . . especially mentioning the traces on the body. These marks are the key to all the rest . . .

Tobie Nathan (to Lucien): Do you know what the father died of?

Bintou: My mother says that I'm the one who "ate" my father.[19]

Lucien: He died a terrible death, then.

Marième (clarifying, for the group): Bintou was born following twins.

Alhassane: Among the Bambara, people are called "Sadjo" if they are born after twins.

Bintou: That's my second name . . .

She started to smile, looking at the floor. I had the feeling that her mind was working fast . . . constructing thoughts, making connections, developing new meanings . . . .

Then we all started to realize what the obvious, but implicit, meaning was. If her mother had previously avoided showing her photos of her father, it was because Bintou was "linked" with him in an imperceptible way and was constantly trying to join him in death. This is why she had fallen badly and the stomach marks came to protect her – to *fix* her – in the world of the living. In addition, coming after twins, Bintou possessed a strange power, hence the accusation of having "eaten her father" (in sorcery). At this point we understood that Bintou had probably been sent overseas by way of protection, to avoid accusations of sorcery against her . . . to shelter her from the gaze of the jealous or envious and, above all to distance her from the morbid attraction she had for her dead father. This is why, despite all of her daughter's material and psychological difficulties, the mother did everything in her power to keep her in France. Yet she remained solicitous and consulted marabouts – Morys, Karamokos or Bammanans – and sent her "things" with which to "wash" (purify or care for herself).

This is where my astute critic comes in to ask: 'Ah, you put it so convincingly! I understand the utility of multiple universes for psychotherapy. I must say that now, as far as I am concerned, and you too, they seem indispensable for triggering the association of ideas for people from Mali. I can also understand how as a team you manage to recreate the ambiance of a village discussion. I also share your interest in getting a group of therapists together to share their knowledge of languages and cultures. But tell me what you think of this question that is nagging at me: how did you deduce that your patient had scarifications on her stomach? You aren't going to tell me that you really are clairvoyant!'

'Look how stuck you are in your ways of thinking. What is important is not what I see, but that I see. I mean, that I make therapeutic statements in this form. Just wait, you are going to be even more surprised by the time we get to the end of this consultation.

Continuation of the consultation

Then I took from my pocket the little black cloth bag in which I keep my shells. I asked Marième to "throw the cowries". She draped herself in her loincloth, spread a carpet on the floor in front of Bintou and crouched down on it. She asked the patient for a coin, mixed it with the first four shells and threw them several times. Then she took one away, then a second, a third and then the last. Then she asked Bintou to throw the shells herself. After that, she gathered up all the cowries, whispered some words in Wolof to them and threw the whole dozen at once, several times.

Then she announced what she saw . . .

Marième: I see a marriage. I see a lot of women talking together, probably arguing, there is also an important man. I can see a lot of argument about a marriage. I can see a lot of wealth as well . . . Bintou will bring a lot of money to a man one day. . . I can also see a woman . . . this woman is sick.

Bintou: Yes! My mother is sick. My brother says she has bad knees but I know it's not true. It's something worse . . . She has something they are hiding from me that they don't want to tell me.

Marième: I see a journey. I can also see that two similar things have to be brought out by making a *sarakh* (offering) of two similar things.

Tobie Nathan: Of two similar things . . . for example a cow and a calf, a ewe and a lamb (*pause*). You are right, Marième, the mama is like a man . . .

Marième: I see the story of a marriage with problems, and

a woman who thinks a lot . . . who has lots of worries
. . .

Tobie Nathan: Well, that's her mother. It's her mother
you are seeing! (*To Bintou*) One day you fell over. (*To
the group*) She was turning, turning, and then she fell
over.

Bintou: You mean . . . while dancing?

Tobie Nathan: Yes, one day, while dancing . . .

Bintou: Yes, I was turning without being able to stop, and
I fell.

And it was Lucien who concluded the divining session!
He said:

Lucien: Things are beginning to fall into place; now we
have to work on consolidating them. A while ago the
family had started applying prohibitions and rules on
Bintou. I wonder if Bintou respects the earlier rules made
for her.

And, as we often do, we finished with a prescription. I
said to Bintou:

Tobie Nathan: When you speak to your mother on the
telephone, tell her you have seen me. Tell her also that I
saw in the cowries that a large animal and its baby have
to be buried alive in the courtyard, and I mean alive,[20] a
cow and a calf, perhaps just a ewe and a lamb. She will
understand . . .

Bintou looked radiant. She didn't even want to leave
the Centre . . . she stayed chatting with the women in
the group . . . .

As soon as I began to introduce her to an invisible world, Bintou began to relax. And we "threw the

Clinical Case: Bintou	One universe societies	Multiple universe societies
Signs/symptoms	• fainting • complaints • somatic • withdrawal • wandering	• absence of a male referent (father, paternal uncle, husband) • stomach scars
Therapists' thinking	• guilty of infanticide/victim of rape or child abuse	• familiarity with death • probable presence of a 'supernatural husband'
Diagnosis/possible divination	• emotional immaturity • split • hysterical defences	• 'father's twin'
'Therapeutic' treatment	• individual interview • medical consultation • psychotherapy	• divination • prescription • sacrifice
Results of the treatment	• 'addiction to the expert' and their thinking • isolation	Renewal of links: • with the mother • with the marabout • with the Bambara family

Figure 1.4   A clinical illustration

cowries" with the aim of formulating a new therapeutic proposition in the form of a prescription. A prescription opens a new signifying matrix at the same time as inscribing it in the real world, the world of things . . . . This prescription sends the patient back to her country, the only place where it is likely to be performed, and it also forces Bintou to talk to her mother. So, in the one act we established relations:

1. between Bintou's suffering and the invisible world;
2. between Bintou and Bambara thought;
3. finally between Bintou and her mother, the only person able to explain to her why we have set such a prescription.

Such therapeutic systems have the curious characteristic of being "contagious", that is, transmitting their effects through contact. So, when Bintou speaks to her mother, the mother will contact the marabout, who will interpret my prescription, an interpretation that will get back to Bintou in another form. In this way, these systems, just by being locked in, install a network of connections and support around the sufferer.

Because of this consultation, Bintou, who was once alone, will now find herself surrounded by her mother, her uncles and aunts, a marabout, perhaps two or three of them . . . QED.

Perhaps my astute critic will find more to say about this. Maybe he will want to know more about how I "fabricate" my prescriptions.'

'I think I have said enough for today . . . another time, perhaps?'

II MEDICINES IN NON-WESTERN CULTURES

Prolegomenas on thought and belief

'I am very happy to be able to pick up on our discussion today. Earlier you asked me how we constructed our prescriptions during an ethnopsychiatric consultation.[21] You saw that the task of explaining it to you gave me trouble. How can one come to terms with precise and complex technical modalities without reconstructing the whole apparatus in its philosophical, theoretical and methodological dimensions . . . ?'

'You are making it sound more difficult than it is. I just want to know if the prescriptions (like doctors' prescriptions) are well thought-through on the basis of sound judgements, or if they are more like an artist's – evidence I must say of your intuition and creativity . . . that's all!'

'Well, let me invite you to participate in an experiment! But first of all you will have to bear with me on a whole series of premises, OK? I am going to try to reconstruct for you the set of elements necessary for understanding how a certain Algerian Kabyle man was cured. He was about fifty-five and we were treating him for alcoholism that he had had for twenty years or so. After he had several workplace accidents, this man was living on welfare, with a basic disability pension. He lived with his wife and ten children in a four-room council high-rise flat. He came to the attention of the welfare department when one evening, dead drunk, he shot at his daughter with a rifle, having mistaken her for a thief . . . .

So this is the situation we will have to explain later on. But before starting my reasoning, I want to ask you to reconsider how you were thinking about two ideas.

The idea of belief
In general, white people think that there are two types of societies – one where people think more than believe and another where they believe more than they think[22] – their own falling naturally into the first category. In order to justify this distinction, all sorts of explanations for the beliefs of others are invoked. How many times have you heard stupidities like this: "Savages in thrall to natural phenomena have invented beliefs to give themselves the illusion of mastering incomprehensible forces of which they are the mere playthings."[23] Allow me, in regard to this, to relate Lorna Marshall's anecdote as told by Mary Douglas in her very astute little volume: "Once when a band of !Kung Bushmen had performed their rain rituals, a small cloud appeared on the horizon, grew and darkened. Then rain fell. But the anthropologists who asked if the Bushmen reckoned the rite had produced the rain, were laughed out of court."'[24]

'I can't understand the meaning of your anecdote at all,' my critic will no doubt reply. 'What are you trying to show? Can one really not say that the Bushmen believe that the ritual makes rain?'

'No!

1. The Bushmen submit themselves to the rain-making rite. This is a reality that is complex, intricate and producing a whole series of actions.
2. And it rains. That is a second reality . . .'

'So, the rain-making ritual has no relationship to the rain?'

'The ritual is a negotiation with powers, with non-humans. It has within it the theory of the world and action on the world. In the end, the rain-making ritual has as little relation to rain as Yoruba scarifications have to leopards, or that the theory of "hippopotamus children" that I spoke about in our earlier discussion has to real hippos.'

'Wow, that seems pretty complicated to me . . . you must admit, common sense is much simpler. So what do you think is to be done with this idea of "belief"?'

'Give it up! Give it up and replace it with a much more reasonable idea: *any specific cultural action – ritual, sacrifice, offering, protection . . . (1) is exactly what it is supposed to be*; *(2) is generally effective on its own terms*. (For example: a rain-making ritual is designed to enter into relations with rain powers, and it generally manages to!)

Therefore such an action does not need interpretation. If this formulation is accepted, then the work of the observer becomes both simpler to carry out and more extensive in its scope. This is because what has to be reconstituted is the totality of the thinking giving rise to the culturally defined action, and this is even the case for non-analytical thought, even if it is thinking via acts
. . . .

Thus, in our discipline, despite numerous warnings

from exceptional thinkers, genuine visionaries (and Marcel Mauss[25] was one of them); despite in-depth and highly revealing research by anthropologists of the "savage mind";[26] despite clinical results regularly published by ethnopsychiatric researchers, psychopathologists continue to make out that there exists *thought* on one side – that of western psychopathology – and *beliefs* on the other – that of (unfortunate) savages who, stuck with fantasies, only know how to naively gesture with "symbolic" acts. And, in addition, more and more relevant research is being carried out by philosophers and sociologists of science that shows that in scientific thought, the experts "believe" as much as the "savages".[27] And yet, due to the construction of their discipline, psychopathologists can only think of "savage" therapies as:

- illusionist traps
- manipulations of belief[28]
- manipulations of seduction
- results of the "placebo effect"[29]
- simple consolations[30]
- "magical thinking", in short (which we take to mean infantile thought)[31]
- the most charitable can see a sort of intuitive science in healers ...

Well, for my part, let me tell you that *other peoples think*![32] I can be more specific. They don't just have real thought, but thought which is radically different from that of "whites". I can add that such thought needs a long and detailed apprenticeship and, just like ours, this apprenticeship can sometimes fail.

The idea of the symbol

As soon as one abandons the belief/thought conceptual binary, one also has to abandon another banality in psychopathological thought, the notion of symbol. Typically, psychopathologists dismiss "savage thought" by pretending to "understand" its symbolism. They think that if "primitive" people use magical methods, it is because they are *symbolically* displacing the deep thoughts that the scientists themselves are able to grasp through "direct contact", that is, without their fantastical veneer. Those who "believe" only have access to the symbol, while those who "think" would be able to access the thing that is symbolized, the very being of the thing.

Thus, one can read that the Wolof wanted to act on the representation of the father, but they *symbolically* displaced it onto that of the ancestor.[33] How can one think for a single instant that, for an African, the father can be the object to symbolize and the ancestor the symbol? How can one invert a hierarchy of values like that when it is endlessly stated and obvious in almost every aspect of daily life? Or, with regard to the Lari or the Bakongo people, that they wanted to attack their own unconscious aggression by symbolically displacing it – "projecting" it – onto the representation of the sorcerer? Banal, wouldn't you say? But when you know what an *n'doki* represents for someone from the Congo, and the terror that they feel at the very idea of being attacked by one of these, one can only be alarmed at the observers' simplicity.

Let me state clearly that it is the notion of "symbol" that gets in the way of grasping non-scientific therapies as real techniques for thinking. This is why I'm suggesting to you a thought experiment where the problem of psychopathological therapeutic techniques systematically avoids any notions of "belief" or "symbol".'

'But, tell me, how are you going to go about penetrating these other thoughts that you yourself describe as radically heterogeneous if your starting point is your own thought, which you agree is of a scientific kind? At a pinch, I can admit that there is some basis to your criticisms. But the methodological difficulty arising from your assertions seems to me insurmountable. Tell me more about your method.'

'I am completely in agreement with what you say. Just to sketch out the scope of the question, shall we try using a seemingly rough method? But you will see that it turns out to be effective. In order to grasp the immense complexity hidden behind the theories of both healers and scientists, I would like you to consider as carefully as possible all of the consequences of their technical activity. Thus we will systematically direct ourselves away from the ideas that the technicians themselves propose to justify their actions. We will systematically avoid discussion of how much "truth" the ideas carry and try to concentrate our attention on the concrete effects of the actions. What do you think? OK, let's begin.

The white man's medicines

Theory
As far as psychopathology goes, whites think that medicines are active, and they pretend they know how they circulate.

1. Psychotropic drugs are synthetic chemicals that are "assimilated" by the organism through receptors.

These receptors are in some way the "sensitive organs" of neurons that themselves are part of the general structure of an organ (the brain) which is the exclusive property of a person. Everyone is endowed with a brain, simply by virtue of being a human being. The use of psychotropic drugs is based on the notion that a person's presenting symptom is caused by a dysfunction of their organ, or at least that acting on that organ will produce some result on the symptom. So far, you will agree, this is just a simple description . . .

2. Now let's describe psychoanalysis. This time, we begin from the other end, that is, the symptom. Psychoanalysis has never paused in its campaign to get people to agree that a symptom is the exclusive property of an individual. It goes so far as to pretend that it is a kind of brand name, as if the whole person were contained, in a condensed form, in a single symptom.[34]

Psychoanalysts coming after Freud – I mean especially French ones – took pains to define the status of this person. Today, they all seem to agree that it is the idea of the subject that best accounts for it: a philosophical "subject" (Lacan, *le sujet du désir*), or even increasingly a legal "subject" (Legendre, *le sujet de la loi*). But this subject is a synthesis of structural elements that themselves constitute a kind of organ, the psychic apparatus. This organ, just like the brain, has sensitive extremities; these are the affects. The instrument designed to establish interactions with the organ – I'm referring to *transference* – always goes via the intermediary of affects. This is why Freud constantly

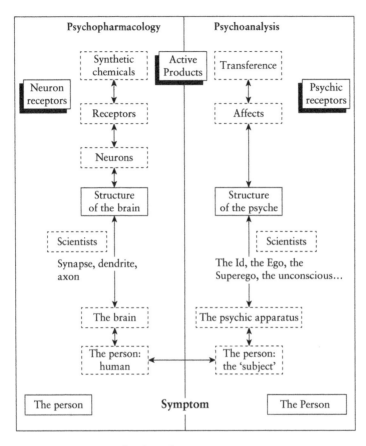

Figure 1.5   Scientific thought

stresses that there can be no psychoanalysis without transference.

I invite you to have a good look at Figure 1.5 and tell me if you are not struck, as I was, by the similarity of the two models of scientific psychopathological thought: psychopharmacology in the left-hand column, psychoanalysis in the right. In both cases, it is a matter of agglomerating (or better, welding) the symptom and the person, then justifying this agglomeration via a kind of logical ascent. Once one has gone from the sign to the symptom (via diagnosis), then one can weld the symptom to the person. And to do this you have to climb through stages: person to structure; structure to receptors (neuronal receptors for psychopharmacology, affects for psychoanalysis); finally receptors for the active products (chemicals for the one, transference for the other).

But the consequence of such a procedure, fairly complex I must admit, is always to *weld the symptom to the person.*'

'I understand perfectly what you are saying. It's just a simple description of the logical sequence that our therapists follow when they go about seeing a little more clearly into our accumulated problems. It seems to me you have a good analysis of the framework of our thought, in particular what you call our "models". But why should this be highlighted? Isn't this just what we call "thinking"? It seems you think that this method is scandalous. Tell me more!'

'I think you haven't yet seen the two catastrophic consequences of the way scientific thought functions:

1. The first is that, by welding the symptom to the person, the person's links to his or her fellows are broken. Once bound to their problem, the person becomes different from their family and friends, mother, father, siblings. She loses, *ipso facto*, the attributes of family, ethnicity and language. She becomes the "object of expertise", sometimes claimed as the expert's property.[35]
2. The second is perfectly obvious. Pharmacologists and psychotherapists can dialogue endlessly, even

if just to criticize each other. They dialogue because they basically understand each other – the symptom should be made as one with the person – they only part company on how elementary parts allowing this mechanism of attribution should be divided (structure of the brain for the former, structure of the psyche for the latter). They think they only need to think to keep such a system going, and at the same time they disqualify all other types of thought.

So, as I was telling you, whites think that they *think*. They even think that theirs is the only way of thinking. When they notice that the others – I am speaking of course of healers coming from non-western societies, blacks – never manage to give up on their spirits, demons, fetishes and sorcerers, they conclude from this that these others *believe*.

But let's pause for a moment on what the whites think that the blacks believe in an opinionated way. According to the first, the second believe:

- that the world is peopled with humans and non-humans;
- that the non-humans, just like humans, are endowed with intentions, and therefore
- that nature is animated;
- that by using these "forces", you can act at a distance;
- that you are not responsible for your own destiny;
- that you can have commerce with non-humans to influence them:
  - through negotiation
  - by opposing them
  - by seducing them

- by supplicating them
- by respecting them
- by tricking them, etc.

Thought is in objects

So, that sums up so-called "scientific" western thought on the subject of psychopathology. With your permission, and in order to follow through the exercise, I am going to ask you a question. What do you think is the most commonly used medicine in the world?'

'I don't know . . . aspirin perhaps?'

'No, it is prayer! And what would you class in second place?'

'I won't be taken for a ride twice . . . Answer it yourself. . . .'

'OK. It is chicken! It is amazing the number of chickens that are killed every day with the sole aim of helping human beings in their suffering.[36] Did you know that? How can it be, despite the fanatical insistence that medical thought should spread over the whole world, that whole populations continue to treat each other by these means? Well, I shall try to explain the reason.[37]

Conceptual objects

You see, the "savage" system finds itself diametrically opposed to the scientific system. What happens when you ask "savage thought" to take charge of a disorder is always a *dissociation of the symptom from the person*. And in order to reach this goal – to break all links that could unite the symptom and the person – all the "savage thinking" that I know feeds back to the same major principle: *attributing intentionality to the invisible*.'

40

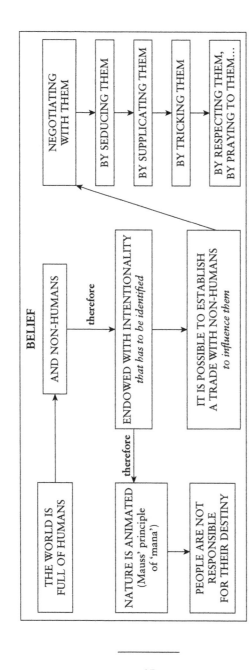

BELIEF

THE WORLD IS FULL OF HUMANS → AND NON-HUMANS

therefore → ENDOWED WITH INTENTIONALITY *that has to be identified*

NATURE IS ANIMATED (Maus' principle of 'mana') *therefore* →

PEOPLE ARE NOT RESPONSIBLE FOR THEIR DESTINY

IT IS POSSIBLE TO ESTABLISH A TRADE WITH NON-HUMANS *to influence them*

NEGOTIATING WITH THEM

BY SEDUCING THEM →

BY SUPPLICATING THEM →

BY TRICKING THEM →

BY RESPECTING THEM, BY PRAYING TO THEM...

Figure 1.6 What the whites think the blacks believe

'May I say, my good friend, that for the first time I am beginning to find you obscure. What on earth is this invisible, and how on earth do you give it intentions?'

'Take the smallest event that creates disorder. Let's say, despite all efforts, that my wife doesn't manage to fall pregnant. We can, both of us, submit to a whole series of biological tests. Suppose we even end up identifying the cause of our infertility: my low sperm count. But that does not resolve the issue. From then on, we need to explain how a fine young woman bursting with good health came to be infatuated with a young man incapable of providing her with children. But she didn't know at the time, you will reply, thinking you are simply making a common-sense remark. Of course! But wasn't there an obscure force that did know? Isn't there an invisible power that even today continues to draw attention to the fact that anyone, like the patriarchs of old, could be chosen as the origin of a new line of descent? Who knows if this sterility, like that of Abraham and Sarah, is not the sign of our destiny as founders of a dynasty?'[38]

'Here you go again, slipping into obscure explanations, and now you are even getting mystical. I cannot countenance this kind of reasoning; you must understand; it is at odds with my kind of rationality.'

'It is time, I think, to remind you of our methodological principles. It is not a matter of discussing the degree of "truth" of the interpretations but of observing what happens when they are activated. If I begin to trot out such an interpretation, I will be bound to interrogate the invisible, wouldn't you say? Identifying the intention it had in creating disorder, in this case my sterility with my wife.'

'You mean to say that all these incredible interpretations that you continually evoke – spirits, devils, demons – are only necessary in order to bring about a technical procedure for interrogation? Is that it? That would be a really revolutionary thought that you have there. Should I be drawing the conclusion that if whole populations are thus required to believe such nonsense, it is with the sole aim of being ready, when a given disorder comes along, to launch an inquiry into the world of invisibility? Wouldn't phenomena relating to belief, in the end, just be kinds of *natural preventions* for the psychological disorders of a people?'

'You are not too far from understanding my thinking, even though you reverted to using the notion of belief, despite my earlier warnings. Give it another try. Let's take a second example: death. If you admit that any event producing disorder is engendered by an invisible intention, then no death can be considered "natural". Well, that's exactly what they say Africans are supposed to think.[39] I would say that in Africa 95 per cent of deaths are attributed to an invisible malevolent being and perhaps only 5 per cent are deaths of the "wise" – I mean of those who have spent their life refining their initiation every day, taking care of their encounter with their double – where death is properly intended. So on the occasion of funerals, you will see most ethnic Africans asking questions of the deceased, who is charged with announcing his own cause of death – generally to reveal who it is who made mischief. As for the wise old folk, it could easily happen that you meet one of them on the verge of death. You could even hear them giving you their exact date of death. And they generally keep their word!'[40]

'Very strange and interesting! Don't you think it weird that the two examples that popped into your head

– infertility and death – are usually treated by somatic rather than psychopathological medicine?'

'I must admit that I deliberately chose those to show you that the divisions we make between medicine of the body and medicine of the soul are only of interest because they contribute to the construction of a discipline. I vastly prefer the notion of "disorder" that leaves open the possibility of inscribing the suffering of the sick into various paradigms. Let me explain!

Considering that any phenomena we relate to "care" can be put into the category that I have roughly designated with the word "disorder", we can now specify three findings that I think have now been clearly established:

1. The principle according to which any event producing disorder reveals an invisible intention is actually a *technical principle.*
2. It is destined to bring about actions. This implies that this principle is not a false theory (a "belief", an empirical intuition, a scientific proto-theory) but a sort of interface between thought and the world. *It is a tool.*
3. But tools are not thinking! Thought is hidden, condensed *into the know-how* of technical activity when it is mobilized, and never, despite appearances, in the statements that seem to you to be so esoteric.

Thus the application of this principle always launches complex sequences associating four elements:

1. the establishment of a disorder;
2. the postulate that the intention is invisible;

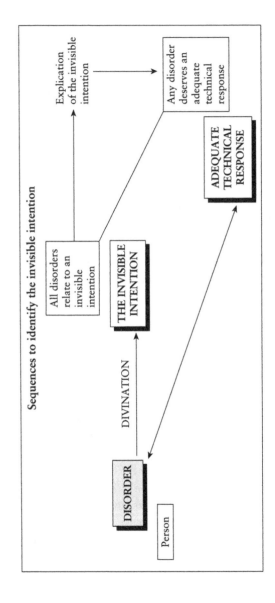

Figure 1.7 Sequences launched by the effort to identify the invisible intention

Sequences to identify the invisible intention

Explication of the invisible intention

Any disorder deserves an adequate technical response

All disorders relate to an invisible intention

THE INVISIBLE INTENTION

ADEQUATE TECHNICAL RESPONSE

DIVINATION

DISORDER

Person

45

3. the explanation of this intention;
4. the most important sequence, often the only sequence visible to an observer: the right response, always addressed to the invisible.

Let's take an example in order to fully understand the progress through these sequences. Suppose the sterile couple I spoke of earlier had been Mandingo from Casamance. Faced with the fact of being sterile, and whatever the other steps taken with (white or "whitened") doctors, this couple would have necessarily ended up with a diviner. Imagine that this diviner consulted the sand (which is, as you no doubt know, a very common technique for interrogating the invisible in West Africa). Imagine that the earth replied that the woman had been married since the age of five with a *djidjinna* (a water spirit) and that, if she didn't give children to humans, then she must have been giving them to the *djidjinnas*, her real in-laws, in a sense. No doubt the diviner would not even take the trouble to explain the whole mythical narrative to the woman, her husband, nor even to the family that had no doubt accompanied them. He would be happy to dispense his prescription. For example, he could ask the woman to sleep alone on Mondays and to dress in white on Thursdays. He could, equally, prescribe the sacrifice of a white lamb. He could ask that the entrails be thrown into the water of the creek and the meat distributed at the mosque or among the poor or to strangers. Finally, he would make an amulet for the young woman, a kind of leather belt which he stuffs with various kinds of plants, perfumes or even writings in Arabic. He could ask her to wear it around her waist until she is pregnant, something he predicts for the following year.[41]

Thus we can concretely reconstitute the different sequences:

1. Report of a disorder – *reception of the complaint*.
2. Premise of the intention of the invisible – *this, in essence, is the very existence of the diviner*.
3. Explanation of this intention – *the diviner questions the sand*.
4. Adequate response addressed to the invisible – *the diviner's prescriptions*.

Popular and expert medicine

Now let's observe the consequences of applying such a principle. If we interrogate something visible – whatever the difficulty (mobilization of competences, sophisticated technologies, complex materials[42]) – we will always obtain expert medicine. Experts, united by the fact of having the same theory in common, will necessarily constitute themselves as a coherent group. Proceeding to question the visible world produces groups of doctors, self-validating groups that can rapidly coalesce into pressure groups. In order to understand what I am saying to you, you have to take a step back. Observe other forms of knowledgeable medicine, Islamic medicine for example (still called "koranic" medicine). It is also a medicine that questions the visible world, that is, koranic texts. It ends up with the constitution of an expert clergy that in turn is constituted as pressure groups, as you know.

So let me suggest the following equation to you. Questioning the visible world involves: the medicine of

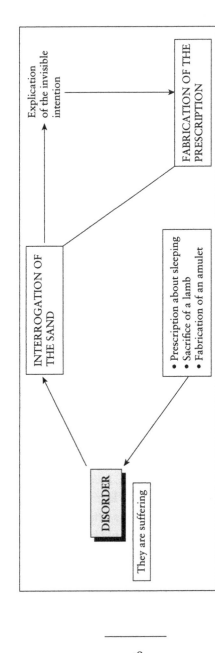

Figure 1.8 Sequences in the identification of the invisible intention – example

experts (the visible is only perceivable by experts); plus the constitution of lay clergy (doctors and academics) or religious clergy (the example of scholarly Islamic medicine).

On the other hand, if you question the invisible world you will always end up giving your patient a new affiliation. The woman in the earlier example will no doubt end up an adept in a congregation of women possessed by water spirits. These women all gather at least once a year during ceremonies that call upon spirits. These are simultaneously: a kind of religious ritual; group therapies; popular festivities; and quite a few other things as well.[43] In any case, and you are with me on this, the application of the technical principle of invisible intentionality always has the end result of creating groups of sufferers or, more accurately, since such groups are already constituted, the affiliation of the sufferer to one of these groups (twins, the possessed, earthly manifestations of ancestors . . .). So, therefore, the first kind of medicine (that of the whites) ends up with psychopathology isolating the sufferer, while on the other hand it reinforces the group of experts (the doctors). The other kind of medicine reinforces the groups of sufferers and isolates the technicians, promoting them by a kind of intrinsic necessity to the status of a singular being.

In the one case, the final result is an expert medicine, a medicine imposed by pressure groups that are always constituted as a clergy of expertise. With this medicine, the peers designate the technician, and always at the end of a long apprenticeship, which actually, as you are well aware, is above all an *affiliation*.[44] No question that he is the one joining a group. With the other, we are dealing with a subtle, popular and always paradoxical medicine. So in a spoken Tunisian dialect they call this kind of medicine *ra'ouani*, which means "upside-down medicine" or "paradoxical medicine". They call it "paradoxical" no doubt because it never addresses what is visible, always relegating that to the rank of a sign. I have seen so many African healers dealing with a clinical problem who don't even glance at the sick person, but bury themselves in contemplation of the sand, water or the entrails of the sacrificed animal . . . Well, in these cases I can tell you that here the "doctor" is designated – or rather I should say "chosen" – by his patients and not by his peers.'[45]

'I have been quiet, not because I agree with everything you say, believe me, but because of the enormity of what you are saying. That you turn your nose up at the effectiveness of modern medicine, the progress it makes day by day, is fair enough – the other medicine basically gets results as well, even if they are, I think, very difficult to evaluate! But your insistence on constantly evoking groups troubles me. Do you mean to say that the wellbeing of populations is squandered by the very people they have charged with delivering it? Could you explain your thinking a little better?'

Modalities for interrogating the invisible
'Wait just a moment and all will become simple. As you have perhaps noticed listening to the brief stories I have been able to put together, questioning the invisible world can only take place according to complex mechanisms – divinations – that are half playful and half sacred.[46] Because the invisible is always situated in substances with indecisive forms in:

- sand[47]
- water
- the sky
- animal blood
- the unpredictable movement of animals, like the call of the *margouillat* (a Reunion island lizard) or the flight of birds ...

And with us:

- in the chance distribution of cards; or
- the shimmering of reflections in a crystal ball.

So let me formulate a preliminary proposal. The idea of divination is not to bring to light something that is visible but hidden, its function is to *establish the very site of the invisible*. If I begin a divination, the technique that I bring to bear presupposes the existence of another universe. Let's take an example. Do you know what a phrase such as "We are not alone in the world" means for the Wolof of Senegal? Don't start thinking that it is about the complaint of a young man in a bad mood. If we reconstituted what is implicit in this sentence, it would give us something like:

> The world is peopled not just with humans. We share it with a multitude of beings such as *rabs* (guardian spirits), *djinnas* (spirits of the bush, waters, mountains, forests, etc.), ancestors ... All these non-humans are there, even if you, as a human, cannot manage to see them. They can brush past you, touch you, call you, appear in your dreams, follow you, hit you ... Some of us manage to see them, sometimes even to begin exchanges with them, you know:

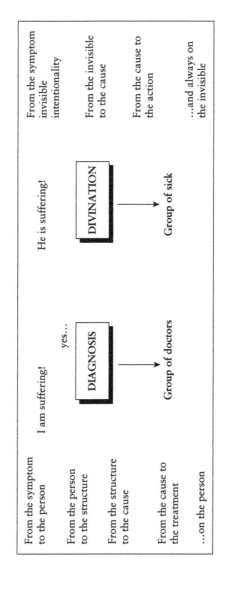

Figure 1.9 Diagnosis or divination

they are children who have not yet acquired speech, mad people and diviners. And by reminding you about all this, by drawing your attention to these facts today, it is because you seem to be complaining about a pain without even asking about where it really comes from . . . So, draw your own conclusion!

That, then, is the sentence with its implicit meanings put back. Its explicit equivalent would then be: "The pain you are endlessly suffering from is due to the vengeance of a *djinna* whose bush plantation you accidently trampled." Strange? This phrase constitutes a reference to the implicit a priori of the group: non-humans exist, their place is defined, as are also the rules for interaction with them! In fact, the initial statement also incorporates an injunction: "Go, then, to consult the specialist on the invisible, the *borom kham kham*, the master of secret knowledge, or maybe the *borom khorom*, the master of the 'hole' (the hidden)."
Because the invisible is in the end only perceptible by a singular being, a being:

- with "four-eyes"
- or whose "eyes are opened" (who has been initiated)
- or maybe who has "an eye in the back of its head"
- or maybe who is equipped with a particular substance in its belly
- someone who is "born old"[48]
- or who is "born with the caul"[49]
- or finally, as we say here, someone "with a gift"

If we stop for a moment to ask the consequence of the functioning of a therapeutic system oriented towards non-humans and towards the clarification of their intentions, then we are led to note that, inversely to our system, theirs produces solitary therapists and sick people in groups.[50] Here, then, is the second formulation:

> The singularity of the one who questions the invisible functions to impede the creation of "clergies" of questioners of the invisible.

African societies seem to think these groups could be in existence, despite all their vigilance. So they are tracked down without pity by types of "secret societies" – like the "leopard men" – groups specialized in putting "sorcerer societies" out of action.[51]

Concepts of the savage mind

> Identification of the 'invisible' whose intention is at the origin of a disorder.

Are you still with me?'

'Lead on! I'll keep a few theoretical questions in reserve. I'll submit them to you later.'

'It seems to me you have a good grasp of the fact that the concepts of the "savage mind" are not found in statements that are made – it is without a doubt that the greatest number of misunderstandings between savage and scientific thought reside here – but in ways of doing things brought about by the application of major principles which are kinds of premises. If you will allow, I am now going to guide you in the explanation of some "savage" premises.

First, since it is a matter of identifying the intention of invisible beings, what do we know about them? For ease of exposition, I have classed them into three broad categories:

I *Supernatural beings – the entities*

- The god of the great monotheistic religions (Islam and Christianity)
- The divinities of polytheistic ethnicities (such as the Yoruba of Benin, the Ewe of Togo, etc.)
- Spirits (guardian spirits of groups or families, spirits of places – rivers, forests, crossroads – spirits of elements – fire, metal, vegetable – etc.)
- Principles (sorts of allegories) – death, respect, family cohesion, sexuality
- Organs (comparison of the functions of the head and the stomach with the Yoruba; liver, heart and breath in Arab medicine, etc.)

Consequence of the technical appeal to entities
It will no doubt seem obvious to you that the appeal to supernatural beings presents several therapeutic advantages.

Being invisible, they are in essence unknowable. Consequently, if the therapeutic process turns out to be ineffective, this will never be a sign of the therapist's incompetence but of the invisible's will to remain thus. So, here, strictly speaking, therapeutic errors don't exist, only tricks played by the invisible ... This no doubt explains why the sick undertake long roads to recovery.[52]

Clinical example
Once we admitted a child of about ten for ethnopsychiatric consultation at the Avicenne hospital. He was

deeply autistic. When his father had noticed his son's absence of speech, he insisted that the activity of a *Rou'hani* was the cause of his suffering.[53] He told us that *Rou'hani* knew how to talk all languages; they could be found talking Arabic, Kabyle and French, of course, but also English and even Chinese. On the other hand, what always came up as a difficulty, was finding . . . a sheikh knowing how to speak the language of this *Rou'hani*, or, more precisely, knowing how to "make him speak".

So, however bad the disorder might be, the principle is always the same: identifying the intention of the "invisible". In such a system, no sufferer is incurable . . . it is the supernatural beings who are sometimes unknowable . . . . The difference is above all one of size, as far as the care of our worst cases is concerned! It is just as the group said: the child is not mute "by nature" or "through illness", because a child cannot be mute. He is carrying out a dialogue, in a secret language, with an invisible being. Find the person most likely to enter into a relationship with this invisible being and you free the child from the dialogue that holds him captive.

2   Sorcerers by nature or innate sorcerers

And yet the recourse to supernatural beings, to the extent that it stresses the unknowable character of non-humans, leaves their intentionality in the dark. The central concept of "sorcery by nature" turns out to be much more active when one gives weight to the second interpretative pole: *intentionality*.

According to this theory, there are beings that look like quite ordinary humans but are equipped with an invisible organ of sorcery. The presence of this "organ" implies a kind of nature. In any case, it can be seen with an "autopsy".[54] Sometimes it is transmitted by the mother, herself a witch, but it can spontaneously appear in a line of descent. This organ affects the sorcerer with well-known types of behaviour:

- he is active at night;
- in an invisible manner;
- according to supernatural procedures;
- *he is said to "eat" his victims*. Actually, he feeds on either their vital substance – the victim can be seen wasting away – or their "work power" – during the night he submits them to a real magical slavery.

In central Africa, this notion is the cornerstone of the therapeutic system. There, caring for a sick person always consists of a "master of the secret" (*n'ganga* in a great number of central African languages) getting the whole family together in which the disorder appeared and then submitting the sorcerer (*n'doki* in Lingala, common language in the Congo, Zaire and Angola) – whom he has previously identified – to a trial by law.[55] As in the American juridical system, explicit confessions of the guilty party have to be obtained. Sometimes these confessions are recorded on cassette tape, sometimes even videotaped. Here are a few fragments of a recorded confession from a Zaire sorcerer (*n'doki*), translated from the Lari language:

> I am going to tell you all the truth about my life. It is my personal story. The life I am going to tell you about is the life I have led . . . a bad life, the life of a thief, a killer, a life where I have done only bad things. And why is that? Because I followed the wrong path, Satan's path. I made magic. I cast spells. I saw all that normal people couldn't see . . .[56]

Then follows the account of his initiation into the world of sorcery under the guidance of his maternal uncle:

> Another time, when my parents were away again, he made me come at 8 in the morning. My parents were in the fields. At his place, he made me sit down. My uncle had a little dog. He took the dog in his arms, and began to wash its face in a bowl of water. He took out the white stuff that is produced from his eyes which he diluted in the water. Then he took a bone of a black cat and soaked it in the water with the rest. Then he ordered me to wash my eyes with this water.

Finally, the story of the cannibal sorcerer:

> My uncle left, leaving me there in the living room; I got impatient and followed him. Opening the curtain a little, without being seen, I spied a human being spread out on the ground. They were cutting him into pieces, like chickens or kids are cut up. His stomach was cut open and the entrails put to one side. His head was cut off and put aside also. They cut ferociously, like one does with an animal.

So we can note that concepts in sorcery are deeply coherent, and give rise to perfectly codified social procedures. As we have just seen, these procedures are taken up by the sorcerers, second actors in the therapeutic drama. The same notion is found in the Maghreb, somewhat attenuated however, but equally widespread under the name of the eye –*'eïn* – usually badly translated as "evil eye". In fact, "evil eye" suggests the meaning that an envious disposition, visible in lascivious eyes, "spoils" the coveted object. But the idea of the eye, on the contrary, brings out strongly a, shall we say, "biological" characteristic of the sorcerer's eye. Thus, it is said that the effects of the "eye" can even affect those one loves the most.[57] In this way, a mother can "cast the eye" on her own child, even though she loves her tenderly.

Logical consequences of theories of sorcery by nature: human sentiments and supernatural procedures
Sorcerers (*n'doki*) are therefore conceived of as beings of human appearance but who seem a little as though they belong to another parasitical species, secretly working at taking over the human world. Their presence is suspected when certain figures are reversed.

- they voluntarily commit transgressive acts like incest and cannibalism;[58]
- they have a tendency to gather in anti-social groupings. It is frequently said that the condition for belonging to a group of sorcerers is to give up one of one's own, for example one's own child, to the cannibalistic drives of the group. Groups of sorcerers are thus "anti-familial groups", "anti-hierarchical hierarchies", etc.

Now let's continue to apply the method that we initially established. Let us ask questions about the consequences of the technical function of a therapeutic system aimed at identifying a sorcerer.

Technical consequences
In such a world, it is impossible to distinguish spontaneously sorcerers from non-sorcerers without the

help of a *n'ganga*. Given that sorcery is conceived as a sort of "biological nature", or even a "character", the most elementary precaution should encourage anyone, faced with a stranger, to heed their susceptibilities. For upsetting a "sorcerer" would inevitably bring about the release of his sorcery substance and the nocturnal consumption of the unwary. So such a system will put emphasis on attention given to the feelings of others. *This type of thought constitutes, so to speak, a psychology* par excellence *since it always requires one to prioritize and imagine the affective functions of others.* And this is not because of "goodwill" but because, and I must stress this, of fear! You see, once again we note how inane the "human feelings" are that we are so proud of, compared to the force of real technical systems totally oriented to giving value to alterity and, moreover, systems designed to require the members of a group to represent alterity within themselves.'

'Wait a minute, now I think you are going too far! You are full of admiration for these systems which, I agree, you describe with a certain talent, but which are essentially running on fear, as you yourself state it! And then you go on to criticize our somewhat more serene systems, oriented mainly towards love.'

'Please forgive the terseness of my reply, but you are being naive again! The systems I am describing are "requirements for thought". Pardon my pessimism, but I really can't manage to go along with invitations – invitations to "love", to be "comprehending", to "listen". For the most part, they are real barriers to thought. But would you mind if we went back to technical questions? One day, a young woman from Zaire told me this story about the therapy she had with a *n'ganga*:

It was the day of my baccalaureate. I went to the examination hall and sat down in front of the question sheet. But the moment I wanted to write, I found my arm was paralyzed. Whatever I did, I couldn't move it a millimetre. It stayed that way for over a year. After having consulted all the doctors, my parents finally decided to go to a *n'ganga* that someone had recommended. First, he said that there was a sorcerer (*n'doki*) in the family. He asked everyone to come to the prayer session. He made everyone speak in turn. He pointed out the sorcerer, who repented. A few weeks later, I was cured.

And when I asked if they had really found the sorcerer who caused her problem, she replied: "Naturally, since I was cured."

In this last sentence, the whole theory of sorcery by nature is summed up:

1. *Premise*: The disorder is caused by the attack (sometimes unconscious because it is nocturnal) of sorcery substance carried by a "human appearance". In this case, as you can imagine from my description, the non-human is hidden in the human.
2. *Technical response*: The *n'ganga* calls in suspects (generally the whole family), interrogates them and, a bit like a detective, identifies the guilty party.
3. *Validation of the procedure*: If, after the guilty party is identified, the sufferer is cured, that means the accusation was correct.'

'But what happens to the poor so-called sorcerer? You are antagonizing me once again, me with my "banal western" aspirations for justice ... Do people go in for witch-hunts after the treatment, like in our Middle Ages? Someone might be "cured" but someone else is "condemned"?'

'That's two questions in one ... First, there were no witch-hunts during our Middle Ages. It was at the beginning of the Renaissance (end of the fifteenth and beginning of the sixteenth centuries) that the western sorcery system began to escalate and get out of hand ... The emerging medical science, coincidentally, had something to do with this destructuring. But that's a problem I don't want to take up here. Let's go to your second question. In my stories from the Congo and Zaire, sorcerers have the choice between two possibilities:

1. Either, as they say over there, they *vomit their sorcery,* that is, agree that their destructive substance be extirpated definitely from their stomachs. The identified sorcerers are then admitted into a kind of internment where they undergo a real initiation, the result of which is that they can, in turn, become *n'ganga.* I have to say that the treatment is enveloped in the most subtle ambiguity because they pretend that the project is about "closing their eyes", but actually they are taught techniques destined to "see" ... sorcerers. In short, witches are transformed into witch-hunters. The people are not really fooled on this count since they are in the habit of saying over there that "Only an *n'doki* is likely to 'see' another *n'doki.*"

2. Or, as Evans-Pritchard remarked in 1937, the sorcerer is asked to swear not to attack the victim any more, in a word, to control his sorcery!

I am obliged out of honesty to note that in certain cases the sorcerer can be badly treated. He can be driven from the village or the district.[59] And only in very exceptional circumstances is he subjected to physical violence, and there are several reasons for this restraint:

1. He is the only one who can undo what he has done. So it is in the community's interest to keep him onside.
2. He is suspected of belonging to a group of peers that might have tendencies for vengeance – a second reason to keep him onside.
3. He is in relations with forces of the invisible, powers of the night. These are relations that can also turn out to be quite useful for the community. To my mind, cases that go wrong are most often in towns where post-therapeutic processes of regulation have not been able to be put in place.

'I don't understand what you mean. Aren't we dealing with a system?'

'Of course! But as far as I know, in Kinshasa or Brazzaville, they have mostly kept the first part of the therapeutic activity – the ways of accusing the sorcerer – and neglected the second, the treatment (or, if you prefer, the initiation) of the sorcerer that can only take place in a village context. Are you feeling better now about the destiny of your sorcerer? Indulge me if I make a couple more epistemological remarks. As you have seen, such a therapy scarcely gets involved with the sufferer at all. In a sense the sickness is an occasion for the community to seek out, find, and eventually initiate an actor who is involved with the most ambiguous and precious things. As it progresses, it weaves complex and profound linkages. Far from expelling X, his symptom plunges him into the deepest secret feelings of Y. In other words, if I present with a certain symptom – ay, depression – it is because my maternal uncle, who happens to be a sorcerer, perhaps without even knowing it, had gotten angry, unconsciously, in my presence. Nice intricate links, wouldn't you say?

Here the questioning of the invisible world progresses via a series of slippages: from the patient's symptom to the identification of the sorcerer, from the sorcerer to the substance of the sorcery, from the substance to the manner of deflecting its effects for the good of the community.

Thus the theories centred on sorcery through nature in their technical procedures:

1. are real family therapies;[60]
2. continually build social cement through the permanent weaving of exceptionally effective interactive mechanisms;
3. require people to ask about humiliations, complaints, lack of respect; in that way, they are a school for social respect;
4. are psychology *par excellence* – or, in any case, what it should be if it really existed: an intrinsic requirement to always envisage the way others' feelings work.

3 Sorcerers by technique: on human (too human) intentions and supernatural procedures

You will have guessed that in this third category the discussion is around voluntary sorcery, what I also call "sorcery by technique". The theory goes like this: You set in motion – mostly despite yourself – a feeling in someone: jealousy, envy, sometimes even love. The harassed other, obsessed by this feeling, goes to see a professional in manipulative sorcery. The latter will fabricate an object designed to calm the excitement of the former, perhaps even to quench their desire. Now this complex object is going to have attributed to it the direct responsibility for the disorder observed. So that, in a nutshell, is a version of the theory that will engender therapeutic techniques. In this case, the intention is simple to guess; it is really always a matter of the same feelings. The guilty party is generally easy to designate; in fact it is often the sufferers themselves who point him or her out among their relatives, neighbours or close acquaintances. Here what is invisible and has to be "undone" is the spell-object. The task of the master of secret knowledge is therefore to ask the divinatory substance about the nature and place of the object. Sometimes he can reveal its presence by way of the divination itself. Hence, certain therapists from the Maghreb are very skilled at making tangled hair (supposedly belonging to the victim of the spell) appear inside the shell of an egg that they have just caused to explode in a fire under the eyes of the suffering person. Would you allow me to tell you another clinical story?

Clinical example

I was treating a woman who had been delirious for about a dozen years. In fact, each time she had given birth, to her five children, she had succumbed to what psychopathologists call "postpartum psychosis". When I met her for the first time, she was wandering at night, running away dishevelled with a baby of a few months of age. She was an Algerian Kabyle,[61] and was constantly accusing her sister-in-law (the wife of her husband's brother) of having made a *s'hur* ("cast a spell"). During the first session, I simply asked her to bring me, next time, an egg that she had kept under her head at night. The night she put the egg under her pillow, she dreamt about an object hidden under the threshold. She woke up her husband, and in the middle of the night he set about dismantling the aluminium draft-protector covering the threshold to their apartment. They found, wrapped in a plastic bag, some Arabic writing in saffron ink. They took this object to a *taleb*[62] without delay, who of course confirmed for them that it was a spell. After this series of events – please believe me – the woman was no longer delirious. I kept the therapy going with her for another two years at least and I can say she got over her difficulties totally. This treatment happened about five years ago. I still get news from time to time. The cure has remained solid until now.

I find this clinical situation particularly edifying, as much because of the way it functions as in its dysfunctions. The woman wandered for years, pleading with family and doctors to sign her up for the sorcery-by-technique system. It is for this reason alone, believe me, that she kept accusing her sister-in-law. It seems to me quite understandable, after all, that the doctors didn't respond to her requests. As you know, psychiatrists hate attributing the cause of a sickness to any kind of exterior object. In a way, it is the very principle of their theoretical operation. But the family also refused her this avenue because she was the only one of her family line in France and was totally dependent on her all-powerful husband – what I mean is, without father, paternal uncle or brother to defend her interests against those of her husband's family. My simple request to bring an egg acted like a trigger for a whole series of events. She no doubt had represented this egg to herself through an image of the eggs of Kabyle seers that she knew of – an egg that can break open and reveal one of the elements of the spell-object. The rest of the events show that in the end she knew that I was incapable of really making the object of the sorcery appear in the divining egg. So she took the task on herself.

Technical consequences

In the world made by such a theory, the main problem is: first, finding the spell-object; second, identifying its nature and its mode of fabrication; then, thirdly, endeavouring to neutralize its effects. So here the questioning of the invisible world moves along another series of slippages, from the patient's symptom to the identification of the evil at hand, from the evil to the spell-object, from the location of the spell-object to the technique of neutralizing its effects – as is always said in these cases, to *undo* it.

The main aim of such a theory is therefore to technologize the therapeutic relationship. One technique created the illness; another technique – or rather a counter-technique – will permit a cure.

The discovery of spell-objects is thus the work of extremely sophisticated specialists. In the Kabyle, you could, for example, be told after a divination that a piece of your scarf, containing a lock of your hair tied in nine places as well as some other ingredients, has been hung in a tree high in the mountains. It continually flutters with the wind. There is no hope of being able to find it. Only the person who cast the spell can lead you there. But as it is known in advance that he will deny any responsibility, any steps that might cause the victim to lose face will be avoided. The solution remaining is to make an anti-spell-object.

To my mind, we find concepts in this theory that yield the most effective of therapies, that is:

- because they allow the hidden intentionality to be made perceptible;
- because they engender sophisticated counter-techniques;
- because they establish complex imbrications of links with the environment, imbrications with two faces: one turned towards the identification of the invisible intention, the other towards the manipulation of the concrete (objects).

Before explaining to you the real functioning of these objects, I will sum up in a chart (Figure 1.10) all the information that I have outlined so far.

You see three categories here (divinities, sorcerers by nature and sorcerers by technique) that for the most part cohabit in these (essentially African) worlds that I have been talking about with you. Of course, some cultures give priority, sometimes explicitly, to one of these categories. As I suggested in passing, West African cultures seem to favour spirits, Central African ones sorcerers, while in North Africa there seems to be a certain predilection for technical sorcery. But look carefully at the figure. These three categories are written into a continuum. And now I might challenge your love of logic again: for the most

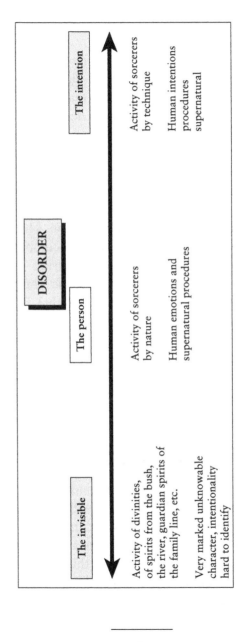

Figure 1.10 Analysis of the intention of invisibles

part, in each therapy, depending on where one begins, *the whole continuum* is traversed. In that way, the one patient can be treated for possession by a river spirit, then for an attack from a "cannibal-sorcerer", and finally for the "tethering" caused by a sorcerer – *ligey*, as the Wolof say. And this is for the simple reason that it is a matter of running through all the facets of the double concept: on the one hand, the invisible, on the other, intentionality.[63] It is as if someone said that in order to really dissociate the symptom from the person, I have to make the two "disarticulating" notions function to full capacity.'

'I am managing, not without some difficulty, to understand what you are trying to tell me. The mythological, communitarian and technical complexity of these ideas would have the aim of signifying – or even more strongly, concretizing – the basic concepts of *invisibility* and *intentionality*. Is that right? But allow me my doubts all the same. Aren't you attributing these systems with a lot more subtlety than they really contain? Couldn't all this be much more simply a case of releasing the person from his or her guilt – perhaps psychoanalysts could even talk of an unconscious guilty feeling – linked to their disorder?'

Active objects[64]

'In order to show you that I am not indulging in a futile exercise of virtuosity, I think you will have to follow me as I set out the techniques. Wasn't my project one of explaining to you how I came to cure a man of alcoholism? Having arrived at this point in this long preamble, maybe I can try out a definition of what a medicine is in a non-occidental culture:

70

In a non-occidental society, a medicine is an active object allowing for the creation, maintenance and finally disjunction of the symptom and the person. A medicine is therefore an object that makes concrete the theory that the community holds on the nature of the disorder.

This definition might interest you because it also allows us to understand what medicines are in our own world:

In our world, in psychiatry, a medicine is also an active object that allows for the creation, maintenance and finally, contrariwise, the conjunction of the symptom and the person. A Largactil (or Thorazine) pill is certainly a medicine because it makes tangible the theory according to which schizophrenia is located inside the subject. In the same way, one can say that a psychoanalytic session is also a medicine.

You can easily understand that in non-occidental societies, medicines are effective only to the extent that they install, and then help maintain, the general theory according to which, at all cost – *and with the help of all available treatment settings* – the symptom must be dissociated from the person. This is obvious in the case of objects such as *protections* – prayers, amulets, sacrifices that to my mind constitute the default medicines in these worlds.[65] That is why I was saying to you earlier that prayer is without doubt the most common medicine in the world.'

'Yes, I now understand what you were trying to say . . .'

'Some other much more complex medicines can only exist because they are antidotes. In other words, this is because, socially, their antagonistic action can be ascribed to something. So there are objects like this that are counter-spells. But from a logical point of view, I should say that spell-objects, such as are used in sorcery attacks, are just as much part of the healing system as are the objects used to defend it.

First, I am going to try to describe the universe inside which objects circulate. What comes first, you will agree, is the technical sorcery attack. A master-sorcerer, lost in the confines of the invisible world – I mean, whose identity is of no interest to anyone[66] – has two sorts of powers: *natural* and *technical*. Perhaps he is himself a sorcerer by nature, by birth. Perhaps he is allied with some spirit or even has a pact with demons. It nevertheless remains the case that he has always learnt techniques. Sometimes it is a relative (uncle, mother) who has transmitted them to him but it can also be a master. Some maintain that it is a non-human who transmitted technical knowledge to them in a dream.

I wanted to warn you right from the start that I dissociate myself entirely from those ethnologists who say that spell-objects, as such, do not exist.[67] Quite often, actually, their presence is only implied. But it does happen that they are actually found. The people who made them are known and often have a reputation. Their effectiveness has been tested over a long time by whole populations, and it is quite real. What seems to me crucial in this whole affair is that these objects belong to the class of things, and not to that of symbols, principles or entities.

What can we call the objects made by the master-sorcerer? Spell-objects, objects of destruction, anti-objects, anti-nature objects? It is still the case that these objects have a certain number of characteristics and, what seems to me fundamental for the coherence of the system, these are the same characteristics that we always find in objects designed to thwart their effects:

- they are always made from a mixture of heterogeneous elements.[68]

  For example: a wooden statuette, containing encrusted metal. If the statuette is X-rayed, a human tooth is found that has been pushed into the middle of it. Animal substances would have been poured all over it, such as broken eggs or the products of several bloody sacrifices, all of which give the wood that recognizable patina. There is also a glass bottle attached containing a certain amount of sand and a liquid. The bottle is tied to the statuette by a piece of cloth coming from some red article of clothing. But we also know that certain words have been spoken over this statuette, and that in the bottle there was a particular perfume that was included but which has since escaped.

- they are almost always composed of elements coming from each of the realms of nature: human, animal, vegetable, mineral.
- they always have an envelope.

  There are different types of envelopes. Sometimes the object is wrapped in an animal skin (fur inside or out), sometimes simply in a piece of cloth. It often needs to be stitched. Sometimes, especially in the case of text written on paper, the envelope is the fold, very precisely made, of the sheet of paper. Some envelopes are more subtle. Perfumes in combination are like this: a certain number of essences are mixed with the use of fire (incense), then a base is impregnated with the combined smell. The base is then composed, for instance in a bottle. The actual envelope is the fire that has brought the odours together. When the bottle is smashed, or simply opened, the envelope is broken. We can thus know for sure that the genie in Aladdin's lamp was the perfume since the opening of the lamp freed him, and afterwards it was impossible to put together an envelope to contain him. One last envelope is even more subtle because its constitution is haphazard. The whole set of elements needed might be brought together and deposited, so to speak, "in bulk", in a pathway, like a crossroads (with three of four paths). It is the foot of some unfortunate that crushes the amalgam that thus constitutes this envelope that has been designed by chance. The "black chicken services" found in Reunion are conceived in this fashion. There it is simply the gaze of the first passer-by that is the envelope of the object.[69]

- they often have a compact kernel.

The kernel, being the soul of the object, as it were, is a subtle body. So most Muslim objects have a "soul" in the form of something written: either the name of God or of one of his attributes. Even the most technically complex objects, like the sorcery statuettes of the Yoruba, Ewe or Luba, or even the Lari bottles from the Congo, are endowed with this kernel, sometimes simply constituted by words spoken during its fabrication.

- they cannot be disassembled.

The envelopes guarantee that the object's character cannot be disassembled because opening them leads to the liberation of the active principle and then makes them uncontrollable. Is it possible, from a simple smell, to decompose each molecule and reconstitute the bodies that went into their fabrication? Of course not! This is why, when you find one of these objects, you run to a master of technical knowledge, the only person likely to be able to "undo" them but not disassemble them.

These five characteristics make living things of these objects. Because in the worlds in which they move, the properties defining the living are precisely: *amalgam of heterogeneous elements; opposition kernel/envelope; possession of a kernel-soul; character that cannot be disassembled.*[70]

There is a consequence to their having the character of living things. Once the objects are fabricated, they pursue their "life" on their own account. They then become independent of the person who made them.

Now, here are the technical characteristics of these objects:

- they are "alive";
- that is, they are active independently of their maker, and naturally of their receiver;
- they are used in an extremely codified manner;
- in the case of remedies, they are "made to measure"; they cannot be given away or lent out;
- all of the thought is contained in the objects themselves; they do not symbolize anything but themselves;

- their fabrication and instructions for use are kept secret;
- if their appearance at a given moment in the development of a society seems very similar from one object to another, it must be borne in mind that their industry is the result of a feverish experimental programme. This is why their form, use, even their names are being constantly modified.

Now I am going to give you a final characteristic, to my mind the most important. Unfortunately, I think you will find it the hardest to accept. The heteroclite amalgam actually contains a *thought requirement*. Above all, these objects are conceptually intricate. They draw their *real* effectiveness from this characteristic. But in order to explain this, I have to take an example. Imagine that in the Kabylia a sorcerer has been engaged to make a *s'hur* by someone with evil designs. This spell is supposed to penetrate the being of the victim to act on her in all her thoughts, actions and attitudes. Now, the object under construction will bring together all of the semantic fields available in this universe concerned with the concept of penetration. In actual fact, such objects, in order to resist really being disassembled, most of the time use several kernel concepts tied together such as *penetrate*, *knot*, *bind*, etc. But to simplify my demonstration, I will limit myself just to the explanation of the paradigm of penetration.

The needle
The first penetration that spontaneously comes to mind is that of pointed objects that perforate the corporeal envelope: an attacker's knife, of course, but also all

those objects that perforate perniciously, such as the needles of pine trees.

Scarification

Scarification is another type of penetration. As you know, scarifications are notches written on the envelope (the skin) in order to signify that there is no definitive closure except death. They can be done on the top of the scalp, on the face (often on the ears in the Kabyle), the stomach or around the sex organs. When they are designed for protection, they are done on joints: wrists, ankles, shoulders. The scar is made indelible by coating the insides of the cut with charcoal, or, especially in the Kabyle, with an ink made from cactus sap, making this a kind of tattoo.

Extractions of 'balls' or 'lumps'

It often happens that through scarification healers "extract" substances that they later show to the sufferers. I have often heard these called "balls" or "lumps", both in West Africa[71] and in the Maghreb. This idea allows for the inversion of scarification, allowing it to go from the paradigm of penetration to that of extraction.

Contrary phrases

There is a range of ideas around certain phrases constructed according to specific rules:

1. They are spoken in a "secret" language, an initiatory or original language.[72]
2. They are made up in such a way that they can't be "disassembled"; that is to say, no one would ever

be able to attribute a communicational intention to them.

3. They have a reputation of "penetrating the person" by throwing them into confusion.[73] They say they "break his head". So they are like the linguistic equivalent of scarification procedures. In Benin, the Yoruba call them *contrary phrases*.[74]

Sacred words

These "contrary phrases" evoke other types of words with "penetrating" qualities. Naturally, sacred words, or prayer words, addressed to a divinity obviously also have an impact on the believer. In the Muslim region, certain litanies (*thikr*), repeated dozens, even hundreds or thousands of times, are reputed to put the worshipper into a trance.

Decomposed words

Sacred words are sometimes "decomposed" through various technical means. In Muslim countries, they can be inscribed in magical squares (*jedwel*) interlaced by knowledgeable accounts:

1. the name of the person in question;[75]
2. the name of his mother;
3. several avatars of the name of God;
4. several avatars of the names of auxiliary spirits, angels or demons.

Words dissolved in water ... or in fire

A very common procedure to envelop words, to make them definitively immune to disassembly, consists in writing them on a sheet of paper and then diluting it in a liquid. The words then spread out as molecules throughout the liquid. After that it is impossible to reconstitute them. A quite similar procedure consists in burning the material on which the words have been written (paper, but also a piece of rag or cloth). The result is ashes in which the totality of the original speech is disseminated.

Infiltration

A person's being can also be penetrated through infiltration, the way a liquid disappears gently into dry ground. In that way, we can say that the words in the paragraph above have "infiltrated" the water, which in turn could then slowly "infiltrate" the person. This is why liquids steeped in active words are used to slowly penetrate people, either in baths[76] or through regular applications that will be left to dry without the help of any kind of prop, such as a handkerchief or a towel.

Fumigation

Another method of penetration, related to the preceding one, is extremely common in Africa. It concerns penetration through fumigation. The utensil associated with these techniques is the brazier on which are placed amalgams of odoriferous plants (lots of resins, of course, but also all sorts of essences); animal substances (mostly fats); and minerals (like alum). The result, a perfumed composite, is mostly inhaled, but it is also supposed to be able to penetrate the pores of the skin. This is why, incidentally, in order to get ready for any kind of fumigation, they take a bath in which they rub themselves with a massage glove.[77]

Unguents

The unguent is a cousin of the preceding procedure. Here, the composite is dissolved in a fatty substance that is used to massage the body extensively. In certain regions, it is common to envelop perfumes in fats. Ethiopia, the Sudan and all of the Horn of Africa, for example, use all kinds of butters to dilute (envelope) fragrances. In West Africa, the same function is attributed to "black soap", in which is imbued not only essences but also substances and, as usual, "active words".

Extraction

As I noted in passing during the discussion of the idea of scarification, penetration necessarily leads to one of its opposites: extraction. Substances and organs included in the same paradigm are "extracted" from the person subjected to penetration. What in fact happens, with the help of the technical object, is the construction of an equivalent to the placenta. This is because the placenta is a quintessential semi-mundane object: (1) half is the property of the foetus that it nourishes, half that of the mother who furnishes it with its basic materials; (2) half to the foetus to which it is tied, half to the environment to which it will be returned. There is a series of human substances that possess these same characteristics of belonging to two worlds: hair (attached to the head but given regularly to the environment); nails; sweat; urine; sperm; menstrual blood.[78]

Here is a list of categories of elements (naturally, not exhaustive) implying an obligation to evoke these concepts:

- Natural elements: thorns, needles, animal teeth or claws, etc. but also, in another register, pigments, spices, alcohols, "penetrating" substances.
- Manufactured elements: perfumes, fats, soaps, honeys, etc.
- Actions that are directly linked to processes of penetration: piercing, cutting, disembowelling, extracting, etc.
- Actions linked to the envelopment of heteroclite objects: heating, boiling, evaporating, burning, diluting, dissolving, enclosing, sewing, binding, etc.

Now, after this presentation (a bit hard to follow, I imagine), look at the table (Figure 1.11) where I have tried to pull together the conceptual field around how "active objects" are made.[79]

Good! Now you will be able to understand how our man from the Kabyle was cured of his alcoholism. We patiently guided him until he found the spell-object. This man actually thought it was his wife attacking him. In the end, he was able to gather traces of the offence: There was a saucepan in which an infusion had been made of pine needles, hair and nail clippings that he recognized as his own, some fragments of paper with Arabic writing, three quarters illegible, but partial verses from the Koran could be distinguished, and certain words changed. But he was still able to read his own and his mother's names. There was also a carefully removed lizard skin, pointed bones from some unknown baby bird and fragments of ashes from burnt alum stone. All this, after simmering in the saucepan, had no doubt been inhaled by the victim, probably while he was asleep.

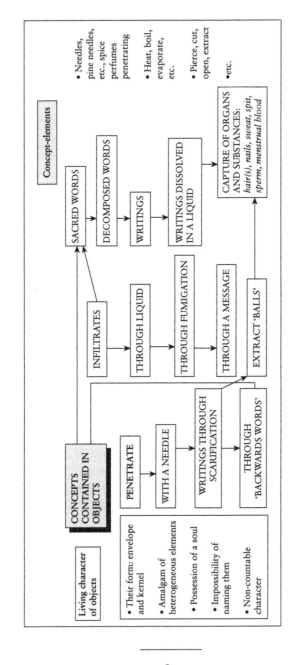

Figure 1.11 Conceptual field of a fabricator of objects

After finding this saucepan, he naturally hurried to show it to us and hand it over (no doubt with a view to having its effects "undone"). But after that he never touched a drop of alcohol again. So, what do we make of that?

We have to agree that this object (visibly a *s'hur*, a "spell-object") was in fact a medicine since it followed the rules of the definition I gave above:

- it revealed invisible intentionality (here the desire of his wife to kill him with drink, or drive him mad, in order to make off with his money for the benefit of her side of the family);
- it disconnected, in a definitive way, the symptom from the man; alcoholism could never be attributed to him again;[80]
- it allowed for the imagination of a technical device likely to construct an adequate response;
- in this case, a counter-sorcery object, a project that he implicitly entrusted us with.

So, my dear friend, my invaluable critic, were you able to follow the explanation? Have I been able to show you that these technical devices are anything but "irrational"? Have I been able to convince you that there are techniques articulated with thoughts that are real, complex, subtle and, above all, socially fruitful, by which I mean that they function to produce intricate relations among people?'

'Well, I have mixed feelings! Your presentation was certainly seductive in its skill. But I still have a host of questions. First, how can it be that you are still the only (or almost the only) psychopathologist who takes seriously these systems that even anthropologists, as far as I know, find a little superannuated, thinking of them as remnants destined to disappear? And then, if all this were true, would we not have to resign ourselves to a complete overhaul of our psychopathology? How can you imagine that happening? Should we abandon our theoretical system – or at least relativize it radically? In short, I don't understand the main aim of your research. Do you simply want to destroy . . . destroy our confidence, our faith in science or medicine? Or are you also harbouring constructive ambitions? Please enlighten me on this point.'

In conclusion

'After having established with you that:

1. there exists in the world an infinite number of effective therapeutic systems;
2. these systems are in no way reducible to ours;
3. they are real conceptual systems and not mere "beliefs",

I am led, following the logic, to rethink what a scientific psychopathology might be, that is, one that is above all respectful of the facts and not just preoccupied with defending an ideology, a doctrine or a type of culture. I now consider that the only object of a *truly scientific psychopathology* should be the most detailed description of *therapists and therapeutic techniques* – never of sufferers. Come on, we should recognize our mistakes! We should forget our symptoms, syndromes, morbid entities, our structures, that are all organized in the defence of a particular clinical approach. No, we can no longer continue looking for mental illness in the ill! I'll say it again: in this domain, all that can be observed are therapists and their objects – and I really mean all their objects: tools, theories, technical thoughts, basic concepts and even, perhaps above all, the supernatural beings put to work by their procedures . . . The work has scarcely begun. The research (it still hasn't happened) should begin with the most detailed analysis of the actors' real techniques[81] working back up to the *theory of these techniques,* before one day turning this into *models of functioning and theoretical objects.*

That's basically what I think. And I can predict that if we proceed as I have suggested, we will quickly realize how many ethnicities, seemingly far removed from what we call "scientific thought", have delivered real conceptualizations. Perhaps we will learn to let them use their own therapeutic resources rather than trying, at any price, to sell them our old stuff, adding violence and abuse for good measure. But what's the use in saying all this? In this domain, how useful is talk? After having listened to me, do you think a single psychopathologist would agree to let go of his or her power as an expert, just to obtain a little more pleasure in thought?'

2
The Doctor and the Charlatan

Isabelle Stengers

Recovering for the wrong reasons

We all know, in fact we are sure, that our medical practices are very different from those in the times of Molière or of Louis XVI. In one way or another, medicine has today become 'modern' along with all the rest of our sciences and practices which call themselves rational. This is obvious, but I would like to question this obviousness, not to debunk it, to show that beyond these appearances nothing has changed, but in order to focus in a slightly clearer way on 'what' has changed. To be even more precise, I would like to focus on 'what' has changed for the doctor, the one who practises medicine.

The set of knowledges being built up today about living organisms, the set of biochemical and metabolic techniques of analysis, modes of visualization and imaging, will play a part in what I want to do but will not occupy a central position because they relegate the doctor to the role of intermediary between the individual patient and a general biological knowledge. Nor will the set of institutions, industries, administrative regulations and financing channels occupy a central position, even though they contribute to the shaping of medical practices. In short, I will not be dealing with medicine in general, with its problems, its inertias, its ambitions, its more or less vicious circles or its occasional uncontrollable waywardness. Nor is there anything sociological about the question I am posing. I am not interested in knowing 'who' the doctor is, but rather 'what' it has meant, ever since medicine became modern, to be a doctor: to be involved with a 'suffering body' and to be involved with it in the context of a supposedly rational framework. In other words, what does it mean for the doctor to be carrying out a rational practice?

Ever since what we call modern sciences came into existence, each knowledge and practice aspiring to rationality must be positioned in relation to this point of reference. Now, from the point of view of the rhetorical and/or practical strategies used for this purpose, the art of healing presents a particular case and that for at least three reasons.

Firstly, this is a practice that could be called immemorial: in all civilizations, among all human groups, in all cultures, there exist and have existed specially designated healers as well as therapeutic knowledges transmitted from generation to generation.

Secondly, the desire to define medicine as a rational practice is, from a historical point of view, fairly independent of the production of a set of practices, which we ourselves deem rational, in the sense of systematically improving the likelihood of the patient being cured. In other words, there is no 'Galileo' of medicine who has simultaneously created a discourse, a practice and a differentiation from the past, forcing us to recognize that in one way or another 'here the history of modern medicine begins'. One might be tempted to slot Pasteur or Koch into this role but they emerge far too late, at a time when everyone thought that modern medicine was already well under way.

Finally, as a profession authorized by a degree, and the product of teaching organized by doctors themselves, medicine precedes by a long way the appearance of modern sciences. In Europe, medicine was taught in the medieval universities and even at that time was locked in a struggle, which continues today, pitting degree-bearing doctors against heterodox or traditional 'healers'. The idea of the regulation of the right to care for patients is a continual feature of the history of western medicine. So at what point does this logic move from a notion of corporate rights, designating medicine as a profession defending its monopoly, to a right which can effectively be authenticated, for us, as ultimately 'rational'? A right which would define the 'real' difference between the practices of modern doctors and those of charlatans? The difference may be invoked on a case-by-case basis but in many cases, even today, it is not always very clear or consensual. I have chosen to construct my approach to so-called modern medicine, starting with the question of its ongoing struggle against those whom doctors nevertheless designate as 'charlatans'. More precisely, I will start with the transformation of the ways in which the 'charlatan' is denounced and the transformation of his or her identity.

The choice of such an approach first of all stems from the fact that, in medicine, the theme of rationality has a polemical accent which it does not have elsewhere. Of course, there is a polemical angle in the way chemistry is distinguished from alchemy, astronomy from astrology and Darwinian biology from the 'static' doctrine of the species. But in each of these cases the polemic is part of the foundation narrative or of an edifying pedagogy. The astrologist is not stalking the astronomer, and the latter does not feel the slightest danger of being confused with this 'other' whom in any case he or she has scarcely ever encountered. There is no procedure in the repertoire of the astronomers which is designed to establish the difference between their knowledge and that of the astrologers. In this case, the polemic is emblematic but it creates no constraint or problem. In the case of medicine, however – and think here of the so-called 'soft' medicines – the danger of charlatanism is quite central. It is continually brought to public attention, explained over and over again in the media and even endorsed by the law (the illegal practice of medicine). Also, as we shall see, it implicitly organizes medical and pharmaceutical research.

This approach also gives us the entertaining possibility of sketching a 'primal scene', that is, something which is both a particular moment and a multifaceted episode where we find gathered, identified and dramatized the matters of concern and dilemmas I am going to draw on for the purposes of recognizing modern medicine.

This scene takes place in Paris in 1784. Two commissions have been appointed to investigate the practices of the Viennese doctor Franz Anton Mesmer. Their main task is to put the principles founding his practices to the test. According to Mesmer, the tub (*baquet*) around which his patients gather concentrates a magnetic fluid, which, through the crises it provokes, has the power to bring about the cures which made him famous. We know that Mesmerian fluid is not part of the present-day therapeutic arsenal, and that therefore it did not survive the enquiry. Nonetheless, we must acknowledge that at the time Mesmer's 'animal magnetism' was a plausible candidate for the foundation of an ultimately scientific kind of medicine. His reference to some unknown fluid, to which only living beings are sensitive, did not disqualify him a priori. It is invisible, certainly, but isn't also Newtonian attraction, whose existence was recognized because of its effects? In this sense, Mesmer's tub could have been recognized as an apparatus which was both therapeutic and demonstrative, its healing power also constituting the proof of the existence of the fluid which explains its effects.

Could have been recognized ... if Mesmer's apparatus had been able to stand the test. I have studied in some detail, with Dr Léon Chertok, the methods of the commission appointed by King Louis XVI.[1] Important scientists of the time, such as Lavoisier and Benjamin Franklin, were named as part of it. To cut a long story short, they tried, without too much success, to 'purify' the phenomena occurring around Deslon's magnetic tub (Mesmer had closed his doors to the inquiry). They then submitted themselves to the 'fluid', then some poor people, then some representatives from respectable society, all without a breakthrough, so the commission invented a much more active method of investigation. It asked an accomplice magnetizer to magnetize subjects susceptible to magnetic crises without warning them; to magnetize them pretending to magnetize another person, and even, the subject having had her eyes blindfolded, to magnetize one part of her body while announcing that he was going to magnetize another. The commission was then able to conclude that 'the fluid is powerless without imagination, while the imagination without the fluid is able to produce the effects that are attributed to the fluid.' In short, the fluid, to the extent that its effects proved its existence, did not exist.

While we bear this scene in mind, let me make a point of one of its features, which is the new definition of 'charlatan' which it carries. In order to explain the cures which nevertheless well and truly happened with Mesmer's tub, the 'interdisciplinary' commission noted: 'We see men attacked, it seems, by the same sickness, cured by following contradictory treatments, and in taking entirely different treatments; Nature is thus powerful enough to support life in spite of poor treatment and able to triumph over both illness and its remedy. If she has this power to resist remedies, then she has all the more reason to have the power to work without them.' Then, the second commission, composed entirely of doctors, raised the stakes: 'There are multiple and sufficient causes for the results said to have been observed in these circumstances: the hopes conceived by the patients, the exercise they carried out every day, the suspension of the remedies which they might have been using previously and of which the quantity is so often harmful in such cases.'

In other words: *The cure proves nothing.* I am going to suggest a definition of modern medicine, in contrast to traditional therapies or to medieval medicine, which does not follow a doctrine or particular practices, which are continually changing in any case, but through the awareness of this fact. It has a correlate: the aim of medicine (curing) is not sufficient to create the difference between rational practice and the practice of the charlatan. The imperative to be rational and the denunciation of the charlatan speak with one voice on this matter. The charlatan is henceforth defined as he who claims his cures are proofs.

This definition of the 'charlatan' makes them modern protagonists as well. Using cures as demonstrations, they make use of a model of scientific truth, rather than a tradition which would imply some 'supernature' which, for its part, would not let itself be paraded for examination on the whim of the curiosity and requirements of humans. It is precisely because the 'fluid' was presented as a 'modern' referent, on a par with Newtonian force, as a 'cause' capable of imposing its own existence on the basis of the examination of its effects, that it could succumb to the critical counter-examination of the commissioners. In other words, not only does the definition of the charlatan I propose not carry any value judgement, since it only functions to define those against whom modern medicine is inventing itself, but also its range is strictly limited. It is Mesmer, not the exorcists whose practice Mesmer believed he had 'secularized' or 'rationalized', who falls under the blows of the critique of the 1784 commissions. The devil would have laughed at the clever trick pulled off by the commissioners.

The commissioners invoke three types of causes to explain the cures attributed to Mesmer's magnetic fluid: Nature's own healing power as evidenced by the spontaneous cures which the living human body is capable of; the patients' trust in Mesmer's treatment; the interruption of other harmful remedies. I will not discuss this third explanation even if it may still be relevant. However, the other two have become a central matter of concern for contemporary medicine. In fact, under the label of the 'placebo effect', the curative power of trust, of hope, and of 'faith healing' are today systematically set out through the protocols which determine the promotion of a chemical formula to the status of a medicine.[2] Modern scientific medicine can thus officially take into account the virtues of 'faith healing', even though it only recognizes it in a negative modality, as a parasitic effect which threatens to impede medical progress if it is not dealt with.

So now we can understand why, unlike astrology, alchemy or creationism in biology, the 'other' of medicine, the charlatan, has not been disqualified once and for all. It is because the charlatan does not just feed off gullibility and ignorance. From the perspective of modern medicine, she or he is the exact correlate of the 'placebo effect' which acts as a parasite in the relationship to be established between a chemical compound and a curative effect. In the same way that the clandestine effect of the placebo must be identified each time, for each new molecule, so too charlatans must be disqualified each time for each new remedy to which they attribute a healing power. This also allows us to understand, in a parallel way, the curious meaning of the term 'irrationality' in medicine. In many doctors' writings, this term is used in order to condemn not only charlatans who use cures as proofs of some kind of snake-oil's effectiveness, but also the public which lets itself be taken in by this proof. Doctors even speak of irrationality in relation to these inexplicable cures themselves, as if, as witnesses to the irrational trust of the sick person, they made this person an accomplice in the creation of an obstacle to the rational progress of medicine.

There is no doubt that we have here a strange use of the notion of irrationality. A priori, particular calculations and decisions can only be called 'irrational' if they claim to be firmly inscribed in the framework of a determined rational enterprise and fall outside of or contradict the guidelines of this enterprise. Now, neither the patient, nor a fortiori the illness he or she suffers, has enrolled in the service of the progress of medicine. It seems to me that we have to understand this usage not only in terms of propaganda – aiming to deflect the public's attention away from alternative medicines or other non-standardized practices – but also in more affective terms. It expresses a real disappointment with this suffering body. When it acts as an accomplice of the charlatans, it provides little or no return on the investment of efforts towards rationality made on its behalf.

So while other modern practices hark back to some original triumph, or to a marvellous narrative about the invention of questions and interpretations which in the end made their object a reliable witness, capable of making the distinction between a scientific statement and a fiction, I suggest that modern medicine has an origin which can be read in terms of frustration: the suffering body is not a reliable witness. It can happen that it will be cured/heal for the 'wrong reasons'.

This frustration awakens old echoes. In 'Plato's Pharmacy',[3] Jacques Derrida brought to our attention the network of characterizations, more technical than metaphorical, which are in play around the term *pharmakon* – poison or remedy – a network which Derrida's reading of Plato associates with the question of writing. Is writing a remedy for memory? In Plato's dialogue *Phaedrus*, this is how Thot, the inventor of writing, presents it to the King of the Gods. But the latter disqualifies it and calls it a poison: 'Things are recollected from the outside, thanks to foreign impressions, and not from the inside of their own accord. So you have discovered a cure, not for memory, but for recollection.' This cure for recollection is a poison for memory and for the soul rendered forgetful for lack of exercise. With due homage to the ambiguity of the *pharmakon* whose effects vacillate and shuttle between remedy and poison, I would like to call the 'Royal Way', the way privileged by the judgement of the King of the Egyptian Gods, a judgement which privilege presupposes a clear distinction: living memory, present to itself, and operating 'from the inside', as against forgetful recollection with its links to prostheses and to foreign impressions. Let me stress that only the soul itself has the power to create the contrast between the spoken word and writing and to distinguish between memory and recollection, and therefore to relegate the so-called remedy to the status of poison. The King limits himself to the role of witness of what the soul truly requires. The Royal Way is not that which the King decides, the King himself speaks in the name of the soul.

Freud can be read as the descendant of the King of the Gods when, disqualifying the specious curative power of what he calls suggestion, he presents psychoanalysis as that which the human psyche truly requires. Analysis does not proceed 'from the outside' via suggestive prostheses, or by a layer of paint applied on the outside (*per via di porre*). The psychoanalyst knows how to get through the surface (*per via di levare*) to the real meaning of the symptoms, without using the least authoritarian prosthesis, without posturing and arm-waving. In this way, it makes itself the witness of the soul, creating a stable disjunction between rational procedure, faithful to the requirements of that which it is addressing, and the *pharmaka* of many unreliable effects, poison-remedies that fail to recognize this requirement.

Some will no doubt think, even today, that psychoanalysis is indeed this 'Royal Way' legitimized by the real needs of the human psyche. I am not among them. That is why the 1784 'scene' (where the commissioners play a trick to disqualify animal magnetism, this early form of hypnosis, the use of which is precisely what Freud was criticizing when he spoke of old techniques of suggestion acting *per via di porre*) inaugurates for me the question which runs through the whole of the modern 'art of healing'. Imagination, which the commissioners attributed with the power to explain the effects Mesmer had set down to the fluid, is equally present in the 'placebo effect' which haunts the pharmaceutical industry, as well as in the suspicion of suggestion which haunts the analytical scene, all the more significant in that it is implicit. Imagination is also present at the heart of the history of psychiatry, where the semiological categories of the 'clinical gaze', which are supposed to decode mental problems, emerge as part and parcel of the uncontrollable mixture which is the historically changing and common matrix shared by the psychiatrist and the patient. The suffering body, or soul, does not have the power to make the distinction the Royal Way suggests: they are not reliable witnesses for identifying the charlatan as he who would illegitimately claim the power to cure.

The power of experimentation

And yet the point will probably be raised that Mesmer's magnetic fluid never existed. The commissioners' method certainly turned the magnetized subjects into reliable witnesses for this partial truth, even if they did not deny the therapeutic efficacy of Mesmer's staging. Also, double-blind placebo trials, which take place on a regular basis wherever a chemical substance aspires to the status of medicine, reliably attribute this status to the substances which emerge triumphant, the ones which have proved that were gifted with a therapeutic power which cannot be reduced to a mere placebo effect. So wouldn't experimentation, then as now, be the 'Royal Way' capable of transforming the materials we have at hand into reliable witnesses?

Quite often, distinctions among fields which relate to modern scientific rationality in one way or another tend to be underestimated. So one might cite the example of astronomy vanquishing astrology, or alchemy made a fool of by chemistry, in order to promote the idea of the same glorious future for all, or to announce that the half-light of situations where the difference between 'rational' and 'irrational' is not clear-cut – is only of a transitory nature, and that all hesitation will evaporate once scientific progress builds up sturdy rational method in each field. From this perspective, the singularity of medical practice would well and truly be in the domain of the 'not yet'.

The history of science does not have the power to condemn an attitude or a hope, but nor does it offer the slightest guarantee for this longed-for triumph of experimental rationality.

In fact, I even think it possible that the successes of modern medicine, remarkable as they are, are not headed in this direction and therefore confront medicine with a *practical challenge*. But in order to explain this clearly, I first need to distinguish the real challenge of experimentation (the one which creates reliable witnesses) from the inoffensive and generalist image it often has of a neutral practice, governed by objective observation and stripped of belief and bias so that it is limited to the establishment of general relations which should in principle give birth to a theory.

There is no doubt that when the commissioners tricked Deslon's subjects, they put the power of experimental method into play. They didn't limit themselves to observation; they actively staged the situation; they invented a manner of setting out the problem of the existence of the fluid such that any parasitical causalities were removed from the scene. Experimentation is an active, inventive practice, and above all selective.[4] It presupposes, implies and makes real the possibility of staging a phenomenon, controlling it and purifying it in such a way that it becomes what it was not, a witness responding reliably to the experimenter's questions. But this possibility, which experimentation brings about, has nothing to do with a method which can be generalized. Phenomena are not subjected to experimentation through the simple exercise of the power to impose questions and extort answers. The phenomenon must be able to satisfy the requirements of experimentation. Experimental purification and manipulation must be validated as what allows it to answer, not as what forces it to answer.

Let me provide what I think is a counter-example, an example of pseudo-experimentation. It is the kind of experimental psychology created by John B. Watson and Burrhus F. Skinner. In order to carry out experiments on rats and pigeons, they invented a laboratory arrangement the purpose of which was the elimination of everything that, in the behaviour of the animal, might bear witness to it being anything but the passive site where two types of observable phenomena had to be articulated – stimulus and response. The description which resulted from this procedure could certainly be called 'objective' since it only picked up observable and quantifiable elements. The upshot was that this method defined itself by transforming into obstacles against objective knowledge the whole set of activities which would make a rat, for example, a meaning-creator, living in an environment which made sense for it. So doing, the imperative for objective description also eliminated everything which may be pertinent for ethological characterization: what makes rats different from other inhabitants of psychology laboratories, from pigeons or humans, for instance? In this sense, the 'objective rat' with its quantifiable behaviour can be defined as an 'artefact', an artificial being stripped of any capacity to answer in its own way to the situations it is confronted with . . . .

So, here, experimentation has not been able to stage a purified behaviour such that it becomes intelligible and capable of bearing witness to its own meaning. Experimentation has cobbled together an artificial, laboratory-created behaviour. It did not endow the rat with the capacity to confirm or refute hypotheses made about it; rather, it 'created a laboratory rat', a rat which is reduced to a mode of existence subject to the imperative of observable, quantifiable objectivity, a rat incapable of teaching us anything about free rats, a rat witness above all to the abuse of power which manufactured it.

Experimentation always runs the risk of creating what experimenters call artefacts: the risk of silencing something while trying to make it speak, of remaining the sole author on stage instead of designing a collective production. Galileo took this kind of risk: he defined friction as merely complicating the movement of falling bodies. The 'real' movement, which corresponds to mathematical intelligibility, is what happens in a vacuum where even air as a source of friction has been eliminated. And this risk was crowned with success. Ever since the beginning of the nineteenth century, engineers, who work in a world where (thankfully as far as our machines are concerned) there is friction, have learned to understand it from the basis in an ideal world described by mechanics, next taking into account the effects of friction which are thus taken to be responsible for the complication of real movement.

Galileo took a risk, and the fact that the movement he was taking satisfied the requirements of the mode of staging he invented, marks what I called an 'event' in *The Invention of Modern Sciences*. The event creates a distinction, which I think is crucial, between theoretico-experimental sciences, which, in every case, have 'made events' and pseudo-sciences, experimental psychology for example, which make the laboratory a place where scientific rationality claims the right to submit whatever it is interrogating to the status of experimental object.

Now, wouldn't it be the case that Pasteur made such an event, an event true to the great lineage which created 'theoretico-experimental' sciences? Did he not take the risk of distinguishing, among epidemic diseases, the question of germs and their propagation from that of the 'field', that is to say the question of knowing how, when in contact with the same germs, some people fall sick or die, and others do not? Should we not acknowledge that in this case we are dealing with a genuine entrance of medicine into the select circle of experimental sciences? Pasteur and Koch were able to isolate germs as the specific causes of specific diseases. They were thus able to 'make these germs speak', to make them perform in such a way that they became reliable witnesses to their own power to cause a disease and to be the vehicle for its transmission. In doing this, they were able to justify a theory of infectious disease, transcending a simple empirical description of the phenomenon. Each epidemic disease had to become the reliable witness for the identification of the germ and the establishment of its responsibility.

Don't these Pasteur and Koch examples show that experimentation is the Royal Way for medicine? The need to have recourse to proof via comparison with a placebo seems to signify, quite simply, that we do 'not yet' have available to us this Royal Way as far as the whole set of our illnesses are concerned. The expression 'rational pharmacology' is a rephrasing of this kind of hope. One day, the curing power of a chemical substance will be able to be deduced from a theoretico-experimental knowledge of the human body, and it is to this knowledge that the substance will owe its status of medicine. For each affliction, one will be able to deduce the appropriate type of action, the structure of the compound with healing power, and it will become less and less necessary to wonder if its effectiveness just hangs on the patient's trust. In other terms, the charlatan, this artist who works the relationships between the susceptibility of the suffering body and 'irrational' influences, will finally meet his or her match. It will be something that can be immediately disqualified since medicine will have the power to act on the 'real causes' of the problem. Both the question of possible placebo effects and the charlatan have a clear part to play in this perspective of a future theoretico-experimental medicine: they are bound to disappear, the placebo because it carries the only empirical character of today's pharmacological research and the charlatan because he or she is the one who will lose, as medicine increases its effectiveness, the power of parasitical seduction. And yet the precedent set by Pastor's scientific triumph does not constitute either a promise of or a first step towards this luminous future. This brings me back to the question of the 'field' which Pasteur dealt with from the point of view of the micro-organisms alone. When it comes to a micro-organism, it makes no difference whether it is in a test tube or a living body. This allows the biologist to characterize this point of view: which environment allows the organism in question to multiply, and which one diminishes its virulence? In the field of immunology, of course, biology and medicine have contributed massively, ever since Pasteur, to the understanding of the field from the point of view of the contaminated organism itself. But this is precisely where any resemblance to the theoretico-experimental sciences disappears.

There is nothing simple or spontaneous about the way in which the theoretico-experimental sciences extend their field of inquiry and the relevance of their practices.[5] Nevertheless, there is a recognizable style to the story which can, a posteriori, set the scene for this extension. A posteriori, what is recounted and transmitted in scientific manuals resembles what the philosopher Kant characterized as the effects of the 'Copernican revolution' which allows a scientific field to depart from empirical practice. For Kant, this so-called 'Copernican revolution' relates to the fact that scientists no longer learn from the phenomenon but impose their own questions on it. This means that scientists hold a point of view on the phenomenon which allows them to determine a priori which questions should be relevant, and which kind of experiment should bring to light the dominant causal relationships organizing all the others.

This is certainly the story one can tell about the extension of Galilean mechanics but not about the kind of medicine derived from Pasteur and Koch. We know now that the question left hanging by Pasteur (that of the epidemic 'field' from the point of view of the reactions of the infected organism) opened a real Pandora's box, and that there is now no story which can, a posteriori, give it a 'Copernican' spin. To the question, 'why does one fall sick?', the immune system, a network where many interdependent causalities are in play, offers no simple answer. Certainly, there has been progress both in terms of knowledge and modes of treatment, but this progress is a long way from keeping up the pace of a mode of explanation which is becoming more economical, more powerful and capable of always establishing more stable differences among what is cause, what is consequence and what is of no importance at all. The definition of the kinds of sickness that begin with micro-organisms has not had the power to become the Royal Way towards the definition of sicknesses from the point of view of the patient's body. This definition has been the entry point into a labyrinth of subtle questions, the ins and outs of which biologists and doctors must explore now and in the future in order to learn from the living body what it is capable of doing.

Pasteur and Koch thought they had discovered the point of view organizing the landscape of the causal or functional relationships defining epidemic illness.

I maintain that, no matter how refined its technical instrumentation, epidemiology is still defined by a form of empiricism: by the necessity to test, observe and describe, in short, to learn from the phenomenon without having the power of deciding a priori what questions to ask it. I stress this point because often the highly technical character of biomedical description is deceptive. How, for example, when we talk of the chemistry of the brain, dare we speak of empiricism while we have more and more detailed images conveying metabolic intensities of different cerebral regions, and while we can identify specific neuronal sites and their corresponding neurotransmitters? Yet it is still alright to speak, even in this case, of basic empirical research. In fact, between the variety of the psychic effects of a drug, for example, and the hypothesis according to which it is modifying the effects of a class of neurotransmitters, there is a gulf which no contemporary theory is able to leap over. What this mode of description puts on stage is first and foremost a set of correlations between two distinct modes of approach to psychic function, two modes which are privileged by the sole fact that they are both accessible to observation.

Of course, nobody would deny that there 'must' be a relationship between the lived effect of a drug and the modification of neurone transmission. But it is exactly this 'there must be a relationship' which defines the practical field of empiricism: the research is dominated by whether or not the observations are accessible, observations among which 'there must be a relationship', and among which all sorts of correlations can in fact be established. But the significance of what is observable, as well as its correlations, is open to an indeterminate number of interpretations. No doubt we have increasingly powerful technical means to measure, and even to create, new possibilities of observation as far as the different aspects of cerebral activity are concerned, but what we measure and observe does not have the power to create agreement about what is thus observed and measured.

Let's go back to Deslon's tub, where what was revealed was the power of experimental staging when it confronts the modern version of the charlatan, the one who makes himself the representative of a 'cause', claiming the power to bring about physiological transformations in whatever circumstances. So what *was* going on around the Mesmerian tub? The tub scene illustrates the asymmetrical nature of experimental staging as appropriated by medicine. It is a negative power, allowing the refutation of claims, but remaining silent on what interested the patients, on the cures which actually happened. Of course, the eliminated fluid was replaced by imagination, but imagination, as well as 'faith healing', is just a way of disqualifying the phenomenon rather than understanding it. Incidentally, Deslon complained that the commissioners did not define what they meant by this 'imagination' to which they attributed the power to cure.

It is interesting to note that among the commissioners, the only one to criticize the judgement of the commission denying all interest in Mesmerian practices was a naturalist, that is, a practitioner of empirical method: the great botanist Jussieu. In a minority report, Jussieu stressed that, even if the commission successfully refuted an incorrect idea, that did not mean it had a greater understanding of what was going on around Deslon's tub. Because the procedure of his majoritarian colleagues depended on a hypothesis of simple causality, all they did was substitute a hypothetically 'simple' cause, imagination, for another simple cause, fluid. In fact, they had staged the phenomenon by defining it as the site of a contest between two possible causes: either fluid or imagination. But why not imagine multiple causality, where 'moral' causes (whence 'imagination') would interfere with 'physical' causes (the action of an agent, as corroborated by the role of the 'touchings' which magnetizers did carry out and which Jussieu himself had occasion to test the effectiveness of)? If the two types of causes could have, in certain circumstances, the same type of effect, the conclusions of the commissioners' inquiry would lose their demonstrative value. Jussieu in fact argued that it was conceivable that the 'moral cause', the idea that one is not magnetized, could thwart the action of the hypothetical 'physical cause', while on the other hand, when the two causes are effectively married, the effect is increased, which is what is observed around the tub. Jussieu finished by calling for an empirical study of the therapeutic possibilities of what he called 'treatments by touch', avoiding grandiose claims and spectacular effects.

Jussieu's objections to the strategies of his experimental colleagues mark the limits of experimentation once one is addressing beings capable of hope and imagination. He did, in fact, stress that 'the idea that one is not magnetized' is not simply the absence of factors linked to the imagination but also brings in a certain type of imagination, perhaps as active as the other and even capable of annulling other effects. The commissioners reduced imagination to a binary variable, which they could bring into play at will, making it act when, for example, they falsely announce to a subject that he is being magnetized, or reducing it to zero when they have a subject magnetized without knowing it. But the imagination does not allow itself to be reduced to zero under experimental conditions. Subjects can't be stopped imagining, interpreting or taking up positions on what they are being subjected to or on what they feel.

Who defines the causes?

As the inquiry of Lavoisier and his colleagues, followed by Pasteur's medicine, showed, experimental procedure indeed constitutes a Royal Way when it puts to the test whether the candidates for causality (the cause of a cure or an illness) have the power to cause by themselves, independently of the circumstances. But this way cannot be adopted by decisions alone (be they Royal, methodological or rational). The king of the Egyptian gods was not able to disqualify writing except by making himself the representative of the soul, which is to suppose that the soul can qualify as a representative. In the same way, experimental procedure requires that what one is dealing with can become capable of manifesting itself through 'experimental facts' which will turn the experimenter into a representative authorized by the fact. This comes about when, in one way or another, the experimenter invents a way of taking the initiative, of staging a situation which responds to his question: the magnetizer's accomplice magnetizing one part of the subject's body, while pretending to magnetize another; Pasteur inoculating some sheep and not others; a doctor medicating in a double-blind protocol, giving some a substance without a physiological effect and others a possibly active substance. Such initiatives signify that the experimenter, faced with a possible 'cause', requires this 'cause' to show its effects in an unambiguous manner, in a situation which has been actively prepared to make a verdict possible. This initiative always takes the form of a variation on the situation, whether this variation is continuous, which is most often the case in physics, or binary (presence/absence), when the staging of the scene relates to logic rather than an activation of quantitative measures.

The fluid invoked by Mesmer was in fact claiming to have the kind of causal power which satisfies the requirements of experimental testing. But it is not the same for the imagination. Imagination is not a true variable because the experimenter is not free to control the variations. He cannot, for example, tell the subjects what they are supposed to be imagining and stop them incorporating 'parasitical' elements which would transform the meaning of the experimental situation. From the experimental point of view, the question of imagination emerges as an obstacle because it constitutes a rival counter-power opposing the experimenter's monopoly on the definition of the therapeutic scene. The living body itself intervenes in the definition of the causes which act on it.

And if the body thus has an initiatory power, if it intervenes instead of submitting, the experimental staging can no longer be defined as the condition for proof. It becomes an irreducible ingredient of the situation. The researcher's initiative (posing the questions, looking for proof) comes up against the fact that the other, the one he is addressing, has not submitted to this initiative, as when a chemical compound is submitted to purification. For this other, the proof is a testing situation, to which he gives a meaning, which affects him along lines which the very conditions of the proof methodology render uncontrollable. The experimental methodology can henceforth become the creator of artefacts, 'facts' which are purely relative to the experimental situation. This, by the way, is what for fifty years the history of experimental attempts to define hypnosis have shown, at their own expense.[6]

Of course, statistical inquiries allow us to circumvent this uncontrollable individual dimension. But between large-scale statistics and the understanding of individual cases, we find the same difference as that which lies between the negative power to eliminate illegitimate candidates to the status of cause and the positive power to understand how 'causes cause'. The first, the negative power, does not lead to the second, the power of understanding, but rather allows us to forget it. And it is when doctors are confronted by this difference, when they are frustratingly reminded of the annoying fact that the living body is an obstacle to the procedures of proof, that they are tempted to speak of irrationality or, with derision, of the 'placebo effect'.

The negative power of experimentation in medicine and the frustration of so many efforts in medical research to positively define the situations it asks questions of are therefore not just simple anecdotal limitations, which will sooner or later be eliminated as knowledge progressively advances. It may be said that the 'question of the imagination' is the symptom of a practical contradiction between the constraints defining the laboratory and the modes of existence of the living creatures who are interrogated there. The laboratory needs a system which will respond to a definition in terms of variables, while the beings about whom the question of the imagination is being asked 'respond' in a very different sense, according to the meaning which they themselves lend to their environment. How can one avoid the artefact if the laboratory must eliminate, in order to give the scientist the power to ask his own questions, the counter-power constituted by 'interpretation' of the situation, whether conscious or not, by the beings interrogated?

Apparently, the meanings which a micro-organism gives to its environment are stable enough so that experimental questioning does not, in this case, create artefacts. This is why Pasteur was able to study the question of the 'field' (test tube or living body) from the point of view of its germs. But the almost paranoid precautions to ensure the reproducibility of experiments in the case of experimental psychology are witness to the fact that, even for rats and pigeons, the experimental staging creates an artefact. It actually creates observable variables (how long does a rat swim in a Porsolt tank before going under?) but the first definition of these variables is that they set the researcher against what he or she is supposed to be studying. Whatever definition we might think of giving to the 'mind' of a rat, one thing is sure: the art of experimental proof carried out in laboratories, where 'animal models' are used to test 'medications' aiming to modify human psychic behaviour, does not put this mind into play but actively denies the problem of its existence. The sinking rat starts behaving normally when it stops swimming.

But in speaking of practical contradictions, isn't one

attributing to the 'mind' of a rat, or that of the patient, something of a spiritual capacity to create their own meanings? Why should we give up hope for a future where this capacity would itself become one of many variables? Simply because in the background the good old opposition between material causation and spiritual freedom continues to assert itself? It is because of this type of objection that it is not vain to consider the example given today by the kinds of sciences in which experimental procedure has worked in the most fruitful manner. It is not a question of looking into these sciences for a 'point of view' on imagination, suffering, interpretation or suggestion. Contemporary physics or chemistry do not offer us interpretative resources. They authorize the simple statement that there is nothing mysterious or spiritualist in supposing that a living body may not satisfy experimental requirements; that there is nothing surprising in encountering 'causes' which cannot be identified as variables, which cannot be identified and put into play at will. In fact, the exploration of the qualitative difference between, on the one hand, systems at thermodynamic equilibrium, or close to it, and those on the other, whose relationships with their environment keep them far from equilibrium, allows us to conclude that it is in exceptional situations where one can identify a cause through its effects in a general and reproducible manner.

Apparently, the difference between these two situations, in equilibrium and far from equilibrium, is purely quantitative and certainly without mystery. In a state of equilibrium, the exchanges between a system and its environment are either nil or balanced, as is the case, for example, when a glass of water is in thermal equilibrium with the room it is in. Maintaining something far from equilibrium simply means that the exchanges with the environment maintain the ongoing character of the system's activity, preventing processes which participate in this activity to evolve towards a situation where they statistically compensate for each other. From the point of view of the definition of the system – that is, the definition of the processes, that of the coupling between them and thus of the mathematical equations that describe them – it did not seem that non-equilibrium should be able to contribute anything new at all. This is why, incidentally, physical chemistry remained for a long time centred on the much simpler study of systems in equilibrium.

Today, we know that this is no longer the case. Far from equilibrium, certain physico-chemical systems are likely to adopt a new kind of behaviour, the behaviour which Ilya Prigogine called 'dissipative structures'.

Dissipative structures introduced a concept into physics that until then had rather belonged to biology or to political thought, 'self-organization'. For the sake of brevity, I will just stress the point that physico-chemical self-organization indicates first and foremost a transformation in the type of causality on the basis of which it is possible to describe the entropy-producing activity of a physico-chemical system. In equilibrium, or in near equilibrium regimes, it is possible to assert that the dissipative activity of a system is entirely determined by its relations with the environment: it is nil in equilibrium; and it corresponds, in near-to-equilibrium situations, to a minimum compatible with the exchanges, which means that it is deducible from, these exchanges. On the other hand, the activity of dissipative structures can no longer be defined as deducible from the exchanges with the environment which are nevertheless its necessary condition. In other words, the 'control variables' which describe the exchanges with the environment here lose their status of sufficient and necessary determinants in order to become constraints which make an activity possible. It is in this sense that this activity could be called 'self-organized'.

The very identity of the system can be transformed in still another way. Factors insignificant at equilibrium, such as the existence of a gravitational field, can come to play a crucial role when the system is maintained far from equilibrium. For instance, without gravitation, whose influence is negligible when a layer of liquid is in equilibrium, the spectacular Bénard cells would not form when this liquid layer is heated from beneath. Far from equilibrium, gravitation is not simply a synonym of 'weight', acting in the same manner on each molecule; it makes possible qualitatively new collective coherent regimes of activity.

The sensitivity of a far-from-equilibrium system to factors that were insignificant, or negligible in equilibrium, is a very important conceptual discovery. In effect, it means that whatever has the status of cause, which should take its place in the description and the prediction of behaviour, is not given once and for all. It is the very activity of the system, which here determines what will, in its case, have the status of a cause and how this cause will cause. Physical chemists usually deduced the possible behaviours of a system on the basis of its definition. They therefore assumed – and this is what we usually mean by 'system' – that defining a system gives the power to define its activity. The notion of physico-chemical self-organization gives us the idea that, in the far-from-equilibrium situation, it is the other way around: activity determines the manner in which the system should be defined.

Of course, physical chemists maintain the notion of system, even in far-from-equilibrium situations. They have the power to do this since what they study is prepared in a laboratory, because the elements in interaction are known to them, because they know what their definition of the system at equilibrium has neglected. The fact that the system can integrate factors irrelevant to equilibrium into its activity thus constitutes for them a new tool for exploration: since the regime of the activity of a system far from equilibrium is not deducible from its equilibrium definition, to study this regime is also to study the stability or instability of this definition, or to ask the question of knowing under what conditions the system can become 'sensitive' to what is, under other conditions, nothing but noise.

Theories coming from physics or chemistry enjoy enormous prestige. This is why I hesitated to use an argument linked to these sciences, fearing it would come back in an inverted form: so that's what the secret of this 'placebo effect' is – a simple question of sensitivity in a far-from-equilibrium situation . . . . Now, the term 'sensitivity' can perhaps keep the precise meaning physical chemistry gives it when it is a matter of posing the problem of phenomena that escape the laboratory definition but respond to the same type of model, atmospheric phenomena for example. The term only functions as an a fortiori type of argument when the laboratory definitions no longer communicate with a practical possibility of exploring the stability of this definition, that is, when the dissymmetry between the positive and negative powers of experimentation is brought to bear. The example of physical chemistry in far-from-equilibrium situations does not have the function of proposing a new model but of dismantling the general view that sees rationality coincide with the triumph of experimentation. There is no need for dramatic oppositions, between the submission of the object and the free agency of the subject, for example, in order to articulate the limits of experimentation. Already the far-from-equilibrium system stops responding to controls in the way that the system in equilibrium responds to them. The way a patient 'responds' to the varied doses of a medication, or the way a dog responds to its human, and the way otherwise quite nice humans, responding to the instructions of an experimenter, may act as executioners in the name of science, as Milgram's experiment demonstrated,[7] raises questions which challenge any generalization. The mode of intelligibility proposed by experimental definition, when it becomes a model for rational inquiry, is a recipe for a systematic production of artefacts.

A practical challenge

Let's go back to the identity of 'modern medicine' as I defined it at the beginning. It might seem clear now that this definition recognizes the 'power of the imagination' but in such a way that the practical questions this power gives rise to are more avoided than elaborated on. More exactly, it implies the hope that one day this challenge will disappear of its own accord, when the dissymmetry that characterizes experimental power in medicine will be reabsorbed; when, with the Royal Way finally opened up, experimentation will be able to identify positively reliable modes of intervention, instead of limiting itself to the elimination of illegitimate means.

Here the famous parable of the lamp post might come to mind. There is a man who looks for his lost keys under a lamp post at night, and a kind passer-by, after trying to help him find them, finally asks him if he is sure he lost them there. 'No', replies the man, 'not at all, but this is the only place where you can see anything.'

Can one 'see' otherwise? Asking this question in terms of a 'practical challenge' means abandoning the perspective of progress that the illumination of new or more powerful lamp posts might symbolize, always extending the field of investigation further. I have not the slightest doubt that lamp posts will multiply in the future, nor the slightest scepticism about the interest in what they will illuminate. But here I want to speak of medicine as the 'art of curing' and take seriously its wish for rationality. This implies putting into the spotlight the situation that prevails today, where rationality is entirely on the side of techniques and of drugs, while the doctor is limited to a representative role, even if it is admitted that her 'human' or 'psychological' qualities create the 'extra touch of soul', both uncontrollable and precious, that characterizes the practice of medicine. Such a situation purely and simply reproduces the dissymmetry that is the hallmark of the negative powers of experimentation: all of the dynamism lies on the side of the accumulation of 'means' which have been tested against the hypothesis that they work independently of the relation between the doctor and the patient, while this relation remains in the shadow of goodwill and difficult-to-communicate experiences.

A practical challenge does not mean a challenge which would be 'only' practical. The term 'practical' has many meanings. Of course, many give it subaltern connotations of the type, 'in theory that is what should happen, but in practice . . .'. But I use the word 'practical' in the sense that all theories presuppose a practice, as well as all our judgements about what exists and what does not. Practice is first of all the manner in which we address ourselves to whatever it is we are dealing with, that is, the requirement that it satisfy certain criteria, and the obligations arising from the way it responds to this mode of address. At the risk of being trivial, I will recall that what we require of a table has little to do with what a specialist in microscopy is looking for in wood fibres, which for their part have little relevance for the techniques of analysing atoms in the chemical sense, which . . . . To be more relevant, I will remind us that Mesmer's 'fluid' does not exist according to the practical criteria of experimentation because it does not satisfy their requirements, but the question that Mesmerian practices evoke is nevertheless still there. They only became 'irrational' retrospectively because they claimed a type of rationality that didn't suit them in the first place.

How can one understand the practical challenge of a 'rational' medicine without going back to the lamp post of experimental progress? In other words, how can doctors become worthy of the problems imposed on them by what they claim to deal with, in this case the suffering body? Experimenters subscribe to the obligations of rationality with which their practice is engaged to the extent that they are obliged to be actively concerned by the difference between experimental fact and artefact. In this sense their practice is, and should be, a polemical one, centred as much on hunting down artefacts as on inventing new types of facts. If the art of curing does not allow one to oppose experimental fact and artefact, if the suffering body cannot become a reliable witness, authenticating the 'real doctor' against the 'charlatan', then wouldn't the 'polemical' definition of medicine, centred on driving away charlatans, become incongruous?

I do not want to suggest by this that the figure of the 'modern charlatan', who believes proof lies in the successful cures he has brought about, should become a positive one. As such, he is of no more interest than the 'placebo effect' itself, both being defined by the question of healing for 'wrong' reasons. Not being reducible to a placebo effect is the test of therapeutic efficacy that the chemical has to pass in order to be considered a medicine, and the charlatan will continue to be considered the 'other' of the doctor as long as the pharmaceutical industry authorizes the doctor to claim that what he prescribes is not just snake oil. But hunting down charlatans has the same limitations as experimentation in medicine in that it allows for the disqualification of the false pretenders and not the positive identification of genuine ones. And it is to these limitations that positive obligations should be attached, thus defining the singularity of the art of healing.

So I will suggest, at my own peril, a radical disjunction between those sites that are no doubt relevant for medicine but do suffice to entail obligations for doctors because what prevails there is the hunt for the modern charlatan and negative proof, and those where, on the contrary, these two polemical imperatives should stop haunting medical practice, sites where it is a matter of curing rather than proving. Whether my proposal is considered 'rational' or not can be verified from the predictable reactions it will provoke: 'But if we give up distinguishing ourselves from these vulgar charlatans,' the doctors will say, 'anything will be permitted; we will be free to do whatever we like!' Such a reaction verifies that what is at stake are obligations that characterize practitioners 'worthy' of their practice, who are not in fear of it becoming arbitrary if they give up referring to the Royal Way along with the fictional idea that the suffering body 'should' become able to tell the difference between the real doctor and the charlatan.

At this point, we should bear in mind that the charlatan, as I have defined this figure, is the modern charlatan. Like doctors, these charlatans consider their activity 'rational' – satisfying the requirement of proof – the distinction being that therapeutic success is taken as sufficient proof. They thus have no direct relation with what I would call, to use the generic term, 'healers'. And here Tobie Nathan's challenging question comes into play: wouldn't we have something to learn from those healers, whose common characteristic is that they are not haunted by the ideal of a Royal Way endowed with the capacity to disqualify others, but rather by having cultivated what one could call, following Nathan, an art of influence.

It is appropriate here to distinguish 'influence', in Nathan's sense, from 'suggestion', 'imagination' or 'placebo effect', because these three terms – even if they are not defined pejoratively in line with the pervasive theme of the irrational – designate an ingredient held to be 'natural', 'psychological', 'found everywhere' and not a technical thought likely to bring specific teaching to the art of curing. Suggestion is what we are all likely to be able to make use of, like Monsieur Jourdain,[8] without even knowing it. Influence implies the expert; it implies a knowledge whose power and interest are, as Nathan shows, to 'technologize the therapeutic relation' (p. 67, this volume).

The way in which Nathan proposes to rehabilitate the 'requirements for thought' (pp. 60, 76, this volume), created by so-called 'traditional' therapeutic apparatuses, a constraint that 'affiliates' the sick person to a world where her life makes sense and in relation to which she can construct herself as a member of a group in which her life is significant, crashes head on into the double idealist register that inhabits us. Equally scandalized would be the two western rivals vying for the Royal Way of therapy: the 'knowing how to listen' of psychoanalysis, and the experimental purification of 'modern' medicine. Using coercion, indoctrination, suggestion and the deliberate creation of artefacts, these two enemy brothers join forces in the terms they use to denounce the treachery constituted by 'the fabrication of brainwashed patients' in relation to the truth project that defines them respectively.[9]

But, at the same time, there is a price to pay for the ideal of the Royal Way. The king of the gods pretended to tell the truth about memory, and the experimental method was produced when, in relation to mute and remote beings, such as rolling balls or micro-organisms, a claim of this type managed to resist any tests likely to challenge it. Western therapies, haunted by the ideal of a Royal Way and by the idea of constituting the suffering body as viable witness of its symptom, are motivated by a requirement that, even if it could never be satisfied, is remarkably operative: they can and must, writes Nathan, 'weld the symptom to the person'. This means that the patient must be 'alone', faced with an apparatus for knowing that defines her or him through a problem whose parameters relate to the collective of experts of which only the therapist is a member.

Tobie Nathan's proposition may be disturbing, but it is anything but irrational. On the contrary, the generation of links and meanings which Nathan associates with so-called 'traditional' therapeutic apparatuses takes up the practical challenge that I have tried to identify: to recognize that what appear to be obstacles to our prevailing ideals and requirements are nothing other than the very singularity of what therapy has to deal with, and to accept the obligation of addressing this singularity without trying to eliminate it or skirt around it. If the psyche, 'the spirit', but also the body, as the 'placebo effect' bears witness to it, are made through relations, they cannot 'respond' to a treatment without also making themselves through their relation with this treatment. Therapeutic influence designates therapists' practical arts, or techniques, of crafting relations which empower a change we name healing.

Having said that, learning to appreciate 'traditional' therapeutic apparatuses does not mean imitating them. As Tobie Nathan often stresses, cultural affiliations are not improvised. If the curer does more than just capture the patient's 'imagination', it is because he too has come from the very culture that he affiliates his patient to. The 'culture' of modern medicine, haunted as it is by the charlatan and the art of proof, defines the sick person a priori as a virtual member of the statistical group that has tested whatever ends up being prescribed or, for the psychoanalyst, as a 'case' who can appear in publication for the edification of his colleagues. This culture is certainly likely to 'affiliate' patients – that is, transform them into living and believing witnesses to modern medicine's powers – and no doubt this affiliation is an ingredient of the therapeutic effectiveness. But, short of founding his own sect, the doctor, as far as 'modern' culture defines him today, cannot aim for such an affiliation or find a way of cultivating it or admit it as an official player on the therapeutic stage.

At this point, I am close to one of the great specialities of western intellectual production, to the very signature of its 'modernity': its claim to indomitable lucidity. It is the lugubrious admission that we have lost forever a resource whose precious nature we only now recognize, but which we are also unable artificially to recreate. So only a grand final movement remains: the appeal to carry out the heroic task of deepening our uniqueness, the disenchantment of the world we are the vectors of, refusing to yield to what is no more than a caricature of what we have destroyed, those proliferating sects that know how to 'affiliate' those who address them. I do not have the least intention of falling into this formulaic lucidity, the great advantage of which is to save those who endorse it from the risk of being labelled naive.

Keeping open a question demands developing it in terms of the 'unknowns' which its usual formulation ignores. Among these unknowns is the kind of apprenticeship which Nathan calls for when he urges us to learn to describe therapists and their techniques in a careful way. The question here is less one of imitation than the transformation of the interested person by her 'object' of interest. Nathan's issue is not just to stop the destruction of others' therapeutic techniques, in accordance with the tolerant maxim 'to each his own technique'. For this type of tolerance, work is not needed. The issue is indissociable from those among us who will have been transformed by this learning, who will invent and be invented by the requirements and the obligations of the therapist's role it suggests. We do not know what kinds of resources they will be able to mobilize in the heart of our tradition or which fragments, which seem to us to have nothing to do with medicine, they will appropriate in order to reconvert them. Did not the Darwinians, for example, redefine techniques of police investigation – traces and clues – in their own way? We don't know. Nevertheless, one thing for me seems certain: it is not just what we call a doctor that will be reinvented but also what we call a patient.

The example of Tobie Nathan himself is very significant on this point: his work has led him to question our so-called civilized society, to denounce not only the way in which we 'care for' migrants and their descendants but also the way in which we, with the help of our standards, facts and good intentions, deny their essential right to maintain the obligations and requirements of their culture.[10] In other words, Tobie Nathan has set out a political problem.

There is nothing accidental or tangential about this. It comes from the living singularity of our tradition that is eclipsed by the very plausible narrative we have inherited about the disenchantment of the world. If there is a tradition that singularizes us, then for me it is the one named 'politics'. Knowing what the city is, who belongs in it and what rights and responsibilities are translated by this belonging, as well as the movements of struggle, inventing new requirements, obligations and matters of concern, are questions which singularize our history in the first instance.

Modern 'rationality' itself is in part linked with this political invention because it was produced first of all as a power of the contestation and transformation of relations of authority and of once dominant modes of legitimation. Today, rationality is not detached from these historical roots: it does not constitute an instance of neutral consensus, overarching conflicts and force relations, but it is an ingredient that itself changes meaning, according to whether it is aligned with the powers that maintain and reproduce the categories through which we define the city or with the social movements that interrogate and destabilize the obviousness of these categories.

I do not intend to transform this text into a political dissertation, so I will limit myself to affirming quite simply that the identification of rationality with disenchantment places the one who utters this statement, whatever her intentions, on the side of the conquerors in our history, those who have known how to capture and suppress its powers of transformation. In a correlative fashion, the practical challenge of inventing for medicine other paths than the Royal Way, which 'weld the symptom to the person', place the one who utters this statement on the side of political invention, which is to say the singular mode according to which, in our culture, minorities invent reasons to dissent and invent themselves.

In our history, it is the *gueux*, the *sans-culottes*, the *plebs*, coming after the slaves, recognizing each other through a Christian god, who were able to invent themselves via the adjective that disqualified them. But isn't this also what is happening now, in the field of medicine, with the so-called 'junkies' who sweep away erudite dissertations about the way to cure them, to reclaim themselves as such in 'non-repentant' users' associations (the Dutch baptized themselves 'junkie' in the act of creating the *junkiebonden*)? I know members of users' organizations are not easy people for the medical profession to talk to because they demand the help they need while refusing to pay the expected price, the recognition that they submit to medical categories. They thus refuse to allow themselves to be 'welded to their symptom'. In my opinion, they are the ones, like organized victims of AIDS trying to get their rights and claims upheld, who are the vectors, no doubt stuttering and sometimes incoherent, of the tradition that singularizes us, the one that we can call ourselves the inheritors of. And this heritage includes those preoccupied with rationality as much as those preoccupied with justice because the 'junkies' who invent themselves as part and parcel of the city, and not as objects of medical and police definition, create in doing this, for us all, citizens, doctors and experts, the constraints and the risks on the basis of which we will be able to work out a discourse on drug use that may at last be 'rational'.[11]

My conclusion, even if it seems paradoxical, was perfectly predictable. It was predictable in so far as 'psychiatry' and 'medicine' are concerned, here as everywhere else, with problems that cannot be separated from what makes collectives exist. In practice, their identities depend quite obviously on the way they accept this inseparability, or define it as an obstacle to a professional practice on the way to respectability. And yet this conclusion is no recipe, 'order word' or denunciation. It does not designate the 'good' and the 'bad', the 'unfortunate alienated sick' and the 'repressive institutions', as if it might be sufficient to throw the latter into question so that the former wake up by some miracle to the possibility of redefining all by themselves what they need and claim. This conclusion has in mind, above all, images that can obstruct current dynamics of invention, or future ones: first, the image that opposes rationality and politics; then the one that would lead a well-meaning doctor to favourably consider self-help movements for drug users or victims of AIDS, confusing them, for instance, with Narcotics Anonymous groups ('associations very useful to keep up the spirits of those involved'); and finally that which would lead to confusing them with associations for sick people 'sticking in a group according to their symptom', which proliferate mostly in the United States. My conclusion aims to propose that the doctor recognize these movements as having vital interest for the future of medicine.

Today, this future cannot be abstracted from such questions. Medicine cannot be reduced to a response to individual suffering and it is not just the business of the doctor and his patient. The way in which humans hope, anticipate, fear and imagine, the way in which they not only conceive but also construct their personal and collective identities, crucially depends on the meaning given by themselves and by others to what affects them. The failed dream of an ultimately 'rational' medicine is now giving way to another kind of rational prospect with the statistical definition of 'risk' groups which directly connect individual ways of life with normative advice bearing on what individuals 'should' do in order to preserve their 'health capital'. In a possible future, rationality may well come to mean a general correlated network of calculation centres making consequential both risk statistics and the way individuals demonstrate their responsibility by their compliance to the norms.

The point is not to denounce rational techniques of calculation in the name of individual freedom. Individuals are not the sovereign authors of the way their life makes sense, neither in our individualist modern societies nor in so-called traditional ones. The point is rather to characterize the way the ideal of rationality, which has defined this ongoing activity of making sense as an obstacle against medical progress, may now come to bless governance – the governing of the flock for its own (statistical) good.

Of course, it is not the doctors' job to decide on questions that are now vastly beyond them. But the terms in which these questions are put will depend nevertheless on the way they position themselves. The dominant position today is that the current development of medicine certainly raises questions for society, but doctors should stick to being modest representatives of a rationality and a vocation that orders them to do what they have to do and demand, wait for or submit to the rules and regulations decided by 'the politicians'. Everyone knows that the situation is not one of such luminous simplicity, but the order-word is still: avoid thinking too much about what would lead to question the categories of what is called progress in medicine, that is to say, stop thinking.

I began by defining modern medicine against the charlatan, and I have arrived at the question of medicine as polity. What sleight of hand was this? One can certainly insist that, from the beginning, the two questions were associated. To return one last time to Mesmer's magnetic fluid, the question that preoccupied the commissioners was political as well as scientific. It is true Mesmer upset not just the medical order but also the order of the city because for many the fluid came to actively symbolize equality among men: the king, as much as his most lowly valet, it was said, could be affected by the fluid, as the 'magnetic relationship' united all humans and affirmed their fundamental equality. Doctors and experimenters as members of the commission privileged the question of the existence of the fluid over the possible therapeutic value of Mesmer's practice, because public order was at stake. But the argument of rationality could be mobilized in the defence of that public order, and this harks back to the singularity of our political tradition. The question of rationality, in medicine or elsewhere, does not belong solely to the landscape of the practices that refer to modern science. It is not only an epistemological question. We are part of a tradition that invented rationality as a political issue and as a discriminating reference for the futures that we are constructing. It has the virtue of a potent *pharmakon*, which can force us to think and, equally, prohibit thinking. When neutral masks of objectivity, good intentions or professional seriousness justify a halt to thought; when the light cast by our lamp posts defines the 'good questions' and actively keeps in the shadows what we might or should pay attention to; then that makes those who refer to rationality and feel authorized by it accomplices to a blind history, which is to say, a criminal one.

That is why the 'practical challenge' of a medicine

capable of becoming worthy of what it is dealing with cannot be understood as a simple local issue, relative to the way a practice, in the drift of history, defines itself in order to be called 'modern'. To become capable of hearing this challenge is also to become capable of recognizing what makes it, whether accepted or ignored, a crucial element of our future.

3
Users: Lobbies or Political Creativity?

Isabelle Stengers

Users: lobbies or political creativity?

I remember an encounter, a long time ago now, in the period in which I was writing 'The Doctor and the Charlatan' (chapter 2). A young woman was telling me about a research project emerging from an experience involving as much her own body as the body of medicine as a whole. She suffered from vertigo, a terrifying experience in which all worldly points of reference are suddenly gone, and you feel yourself falling. Such an experience has its own imperative: go to see the doctor.

Is another kind of medicine possible?

But for the doctor – and who can blame her? – vertigo is first of all a symptom, perhaps a symptom for lots of things, from a tumour to some unidentifiable X, so for her it is a case of working back as quickly as possible to whatever the cause might be, that is, to make the symptom stick to some category that points to an adequate treatment. If there is one. If there isn't, then the problem is that a somewhat fatalistic compassion replaces the doctor's active involvement. And, for the patient, a kind of shame begins to creep in as she feels that she is imposing on doctor after doctor with this 'difficult case' for which they have no answer. Then one day one of them makes a passing remark about a group of people with the same complaint, brought together by the need to generate knowledge that is not about vertigo as a symptom, but about the terrifying experience that has to be anticipated, managed and monitored – in short, that has to be civilized. The young woman's project was to join this group and work with them as both one of them and as a researcher.

I never saw this young woman again, but I never forgot her. She was the fleeting precursor of a possible future. I certainly knew that sick people sometimes get together in groups, and that doctors are prepared to collaborate with these groups. But what sparked my imagination was the idea that knowledge produced by such collectives had a chance of moving out of the domain of 'the anecdotal'. I am very sceptical of the kind of interdisciplinarity that cobbles together initially disconnected kinds of knowledge, as in the expression, 'psychosomatic'. But the possibility was there that such collectives, coming together over a common affliction, could produce types of non-medical knowledge, each and every one in its own way. And the possibility that the knowledge produced could be communicated in its own right opened up totally new perspectives that could conceivably modify medical practice itself.

The collective process of transformation of the powerless complaint of an isolated being, dependent on a knowledge for which the categories are developed somewhere else, into a point of view articulated by the very ones who need it belongs to the order of the event. It is an event of political creation. Such events are decisive in the history of types of knowledge pertaining to humans. It happened for instance when women, learning to become feminists, started thinking together, and when their personal suffering was transformed into active, political understanding ('the personal is political'). The point is not that feminist readings are 'true' in the sense they can be checked against the way things are. Their truth is in the transformation they induce. As soon as that happens, no researcher can continue to allow himself or herself to calmly accept the way in which their predecessors used to characterize 'women'. The same thing happened when addicts founded mutual-support associations and demanded to be recognized as 'citizens like any others', rather than either as delinquents to be punished or as sick people in need of psychological help. They produced an analysis of the situation that modified the way in which experts in drug addiction defined the problems associated with illicit drug use. Of course, certain consumers of drugs need psychological help; it is just that this need doesn't define the problem posed by drugs.[1]

In the case of 'vertigo', it would certainly not be a matter of claiming one's status, nor battling against something, but 'just' a matter of constructing. Constructing knowledge with words and practices that do not deny vertigo as a symptom for the doctor also makes it into an experience that is significant in itself. And this process of construction, if it were to be recognized as such, cultivated, narrated and developed with researchers who agree to run the risk of first finding the appropriate position (neither voyeurs, judges or neutral reporters), could possibly become of interest for doctors as well as a relevant part of their training. Because such knowledges, once they are articulated, can be echoed by doctors, who would no longer be dependent on avenues of treatment alone. Their compassion could be transformed into active intelligence, openly indebted to those through whom it was nourished, creating a link between the 'patient' and the words created by others

---

who are equally 'affected', orienting her, if possible, towards groups where what she is living through can be constructed, where she can become capable of thinking through and experimenting with what is happening to her, not just submitting to it.

Let us not deceive ourselves. The language that might emerge from such groups would not be the authentic 'lived truth', in opposition to the 'technical', 'dehumanized' competence of the doctor dealing with what, for her, is a symptom. A constructed knowledge is never authentic; that is not what is required of it. It is therefore a case of a constructed, operative knowledge creating a distance from the 'lived', as medical knowledge does. But created distances bring into being different practical spaces, defining the 'illness' as their common referent but via different modalities. The groups coming together around what 'affects' them would not just be mutual-support groups, sharing their suffering and advice, but groups producing a 'knowledge that counts', 'user associations', so to speak, which give the object that factually unites them the power to make them think. And since each time knowledge is born in the place where there was a complaint, this would also be a 'political creation', which requires the so-called rational approach to situate itself, which requires the doctor to speak to his 'patient' not just as someone to whom he has to listen politely, someone from whom he can learn nothing.

Thus was sketched out the imagination of another possible medicine, rather than the vague dream of a 'more humane medicine'. Medical studies that could initiate students in multiple languages, all coming from elsewhere, some no doubt from the background of pharmaceutical research but others from concerned groups, would pique the interest of these future doctors about what their role would be concerning the connections among these diverse but equally needed spaces. Doctors trained in this way would not just read medical journals talking about new drugs or diagnostic techniques but also publications reporting on learning processes which are just as relevant for them.

I was, of course, naive, because I didn't know at the time that there were others also taking an interest in 'user associations', though in a somewhat different way. These were lobbies linked to what Anglo-Saxons have called *disease mongering*. So one could say that what I had thought was imagination had turned back into a dream, and in this case a dream deflated like a balloon coming up against 'reality'. I prefer to go further in order to try to learn on the basis of what has happened.

Disease mongering

Mongering means both selling and promoting. We call war-mongers those who promote wars, usually to sell arms. In April 2006, the online journal *PloS Medicine*[2] published a special issue on disease mongering, defining it as 'trying to convince essentially healthy people that they are sick, or slightly sick people that they are very ill'.[3] The procedure is characterized by several stages: put in place a set of symptomatic criteria, each one of which may have multiple significations but which, together, define a disease, that is, something that should be treated; let it be known that a good sector of the population suffers from this illness; let it be known that it is a matter of a deficiency or something off-balance that can be corrected by an active chemical; present the treatment as without risks (especially in the long term); finally, if necessary, make selective use of statistics to exaggerate the benefits of the treatment . . . The success of the operation lies in the creation of groups united by the cause that constitutes this 'new disease', thus claiming that it be recognized, that doctors learn to diagnose it and that its treatment be covered by insurers.

Disease mongering is not limited to the 'psych'-related fields, but I will concentrate on them, as they are especially vulnerable to mongering operations.[4] In fact, the definition of a set of symptomatic criteria permitting the diagnosis of a 'disorder' is the *DSM*'s accepted procedure. One could go further to affirm that ever since psychiatry has been in existence, including psychoanalysis, all diagnoses correspond to the success of a 'sale', and that certain sales have been pulled off through real lobbying work including patients (I will not even mention psychoanalysis or remind the reader of the multiple personality epidemic in the United States). Nothing new about this phenomenon, then.

Things are most often like this in human history. It is always possible to 'dramatize' something new, or relativize it in the manner of 'actually, it isn't as new as all that'. Those enamoured of critique will pause on this equivocation and snigger. But, when they do, the critique situates itself outside of the problem in terms of which the question of novelty may or may not be interesting.

If I wanted to fight against the idea that 'good psychiatry' is possible and that its possibility should be defended from nasty disorder-sellers, I would adopt the 'this isn't so new' position. But the thesis I'd like to uphold implies that the new name, disease mongering, is actually translating a new problem. Let us be clear: this thesis is not aiming to tell the 'truth' but to construct the problem in such a way as to escape both the sniggering relativist conclusion and the call to defend 'real medicine' from impressionable users getting involved in something that should not concern them.

In fact, I have a double thesis. It is first a question of bringing to light what is effectively a new dimension of the phenomenon. While in the past psych specialists would have been deeply offended by the accusation that they 'sold illnesses', what we are dealing with today is actually organized around actors for whom selling is the official problem, that is, pharmaceutical industries. Correspondingly, the question is no longer one of a critical position (that has become too easy) but one of the position to take when confronting a process where actors are redefined, a process which, as such, has the capacity to redefine the whole psychotherapeutic field. Secondly, it is a matter of speculating on the way in which user associations might play a crucial role in the matter of disease mongering, if they succeed in changing the problem, in this instance, in 'demoralizing' the question. I must stress 'in the matter of disease mongering', meaning that what is at stake in this case is not just characterized by competing diagnostics, categories and etiologies in relation to a complaint recognized as needing, in any case, effective help but is first characterized by the production of the 'true disorder' category at the point where the people involved had not previously thought of themselves as 'sick'.

A machine

In order to emphasize the novelty of the disease-mongering process, I shall describe it in terms of a new 'mechanical assemblage' in Félix Guattari's sense.[5] It would well and truly be a machine that emerges and starts to conquer and reconfigure not only the 'psych' field but also a set of questions usually related to ethical, political, cultural and social problematics.

The appearance of a new machine, in Guattari's sense, implies that elements that were previously more or less independent, or interrelated in diverse fashions, are 'grasped' together and begin to work together in such a way that henceforth it is on the basis of the mechanical functioning itself that they have to be described. The machine is like a new subject; the elements have become 'its' parts. In this sense, the engagement of a new type of mechanical functioning is an event, the coming into existence of a new being in the heart of the environment that this being transforms and annexes into the parts of its own functioning.

An important characteristic of the event of a mechanical 'grasp' is the disarray and/or impotence affecting the critical position. What could have, until now, been effectively criticized as weak or even laughable stops being so. Each element of the machine remains easy to denounce, but once this machine starts functioning, it is the very weaknesses of its different isolated parts that gives it strength. Each needs the others, and they each owe their role to the functioning of the machine itself. In short, what was weakness becomes strength, and the machine couldn't care less about criticisms. We have already seen an example of this: former 'psy' practitioners might have been affronted, and therefore disturbed, by the idea that they practised disease mongering, while the pharmaceutical industry will no doubt complain that they are sincere about 'working in the patients' interests', but it will not be disturbed: selling illnesses is part of their standard publicity operations, just business as usual.

Weakness becomes strength. One can, along with Philippe Pignarre, associate 'psychotropic' medication treatments claiming a scientific approach to neuronal functions with 'petty biology' and 'petty psychology'.[6] Petty biology is where a chemical is statistically associated with 'effects' on behaviour, and any supplementary account giving a causal interpretation to the statistical association is deceitful. Petty psychology is where categories identifying the 'disorders' are defined and redefined according to the effects of the chemical – not all its effects, only those that have been able to be statistically associated with it during clinical trials. Now, the weakness of what parades as biology and psychology in the matter of medication treatment of 'psych' problems is not, from the point of view of mechanical functioning, a problem. On the contrary, it is a condition of its functioning. Without this weakness, it would be impossible for it to annex new territories or to transform new sectors of the population into 'markets' for a new chemical because the new chemicals put on the market should then prove themselves in a 'stable' landscape whose contours can't be modified according to the opportunities.

In the same way, one can criticize the questionnaires that allow for the identification of the disorders in the *DSM* repertoire and show, for example, that the patients very quickly understand how they need to reply if they want to enter one or another category or to avoid it. But this weakness becomes a strength from the point of view of the 'sale' of the illness, precisely because the questionnaires have an overtly inductive value, transforming into diagnostic elements problems that had not, until then, been considered relevant to the 'psych' domain. These questionnaires form a part of active recruitment operations. They are sent out for self-diagnosis purposes, accompanied by a pitch: we are dealing with a newly identified disorder here; those suffering from it should inform their doctor, specifying, as the case may be, that he prescribe the appropriate chemical.

These questionnaires promoted for self-diagnostic purposes, or for parental diagnosis, are somewhat analogous to the little quizzes that can be found in magazines for adolescents – are you the jealous type? Passionate? A fabulous lover, etc.? But they contain the key recruitment message: 'Be aware that what you are suffering from may be a real disease!', not just a personal problem that you should somehow deal with yourself. And at the same time this message sets in motion and interconnects new patients, journalists, doctors, politicians, etc.

Internet sites, in this context, are therefore not simple information vehicles. It is no coincidence that pharmaceutical industries subsidize them the way they subsidize scientific conventions. These are sites often created by associations of people recruited by the new sickness, addressing others who are 'still suffering' for no other reason than their ignorance or because of their doctor's incompetence, not knowing that what they are suffering from can be cured. These sites function as concentrators and producers of identity and competence. 'Visit the CHADD[7] site, then you will be able to explain to your doctor. . . .'

The message 'It's a real disease!' that the questionnaire activates functions as an operator for capture and link creation. Henceforth linked and held together by the questionnaire are the new patients and the pharmaceutical industry, and also the psychiatrists and the neuro-cognitivist researchers attracted by the perspective of scientific progress. It is therefore the questionnaire itself that produces the actors, mobilizing them around the disorder as a cause – no one should suffer any more, through ignorance, from what we know now is a 'real disease', and a curable one. The identification of the disease is celebrated as the triumph of science over belief: we 'believed' that it was stress brought on by the cult of performance or perfection, or poor self-image, lack of confidence, or . . . . Now we know.

The case of so-called hyperactivity disorder is exemplary from this point of view. It became an attention deficit disorder, and from then on was interpreted as affecting the 'executive command centre', and it is invoked by cognitive theorists as evidence for the existence of such a 'centre'. What, until now, made the lives of parents and teachers a misery now interests those who deliver us a 'scientific at last' conception of the brain.

The moral connotations, which bear as much on the parents who don't know how to bring up their children as on the children who don't know how to pay attention, listen to what is being said to them or understand that one shouldn't disturb others, disappear in favour of an 'objective' redefinition founded on the effectiveness of the chemical and promising a better understanding of cerebral mechanisms. Grand narratives of the progress type can then build up. While before teachers could be blamed for impatience, parents could be made to feel guilty and victims could be preached at, now we know that it is nobody's fault, and people who had been blamed are vindicated and henceforth called upon to play roles of witness and expert in the diagnosis.

Condemnation?

We can speak of lobbies where there are associations of people who have been recruited by the machine, who propose edifying stories of their journeys of suffering before they were recognized as 'sick' and of going on crusades so that unnecessary suffering can be avoided for those who can be treated, so that the disorder can be detected, so that doctors can be forewarned, even put under pressure and so that insurance companies pay for the saving treatment. But be careful! Speaking of lobbies is not the same thing as condemning them.

As Philippe Pignarre stressed with regard to the question of the epidemic of depression diagnoses, it is important to avoid a reductive critique which would turn a deficit of attention, for example, into a 'simple' social construction, 'simple' meaning hiding the true culprit, liable to deconstruction, for example our society's intolerance of any agitated child or else parents who can't manage to play the role of parents, escaping their responsibility through medicalization. I want to emphasize that the guilt loaded onto parents whose so-called badly brought-up children are destroying family life and are set up for scholarly failure *is also a construction.* And, besides, even those who are against medicalization have to acknowledge that Ritalin can have beneficial effects.

So the question is not construction or falsehood; it is, rather, the way the issue is mobilized. And, more precisely (because multiple personalities have also been subject to a similar mobilization when no drug therapy existed), it is the process of production through repetition of such mobilization, nurtured by the permanent flux of the pharmaceutical industry's production of chemicals searching for diseases and disorders. The industry does this through continual research that associates chemicals with a complaint which could become a clue for a 'disorder'. The slightest correlation, often discovered by chance, and the cogs of the machine start to turn, fabricating both the medicine, the disorder corresponding to it, an evaluative scale that permits the recruitment of interested parties and the recruitment of lobbies insisting that no one should have to suffer needlessly and that troubles don't have to be accepted with resignation or courage because 'we now know' they are symptoms of the new disorder.

The name given to the process, disease mongering, quite obviously has a part to play when it comes to condemnations, sounding warnings and making accusations. And the pharmaceutical industry is the first to be accused. And yet this condemnation has the weakness of leaving unquestioned the way protagonists are enrolled and their roles distributed. It can attack each part of what I have described as a 'machine', and each, as I said, is actually 'weak', vulnerable to condemnation.But each is captured by the mechanical functioning and can point to other parts as what is 'making it do' what it does. Even the pharmaceutical industry can say: 'It's not our fault; the patient associations would not understand if we failed to help them distribute relevant information.'

Condemnation is hugely tempting. But it is not in the name of truth that I claim this temptation should be resisted. It is because giving in to it might give new force and actuality to another 'construction', that of the exclusive authority of the medical body where the definition of real diseases or disorders is concerned.

Condemning the pharmaceutical industry for trying to make money by all means possible will not go very far: that is its role. On the other hand, the novelty of the lobbies mobilized in its service to gain recognition for a new disorder, or its seriousness, offers a tempting and vulnerable target. But what will be denounced then is the fact that incompetent people are getting involved in something that might be of concern to them but is definitely not their business. And the conclusion we are in danger of arriving at is that the definition of disease and disorder should be left in the hands of the practitioners!

As we have seen, learning to get involved in something that is not supposed to be one's business is precisely what I define as a 'political' event as exemplified by the appearance of users' associations which transformed the landscape of authorized expertise, showing that it was defined not only by the knowledges in place but also by the exclusion of those that were defined as incapable of producing knowledge that might count. Patient lobbies recruited by *disease-mongering* operations can thus provide ammunition against such political events. The defenders of the authority of expertise will rejoice in this confirmation of the need to protect it against people ready to accept the promises of miracle workers.

So it is politically important to construct differences between patient associations mobilized by the promotion of their newly recognized illness and users' associations producing actual knowledge in relation to the landscape where they are situated, involving diagnostics, treatments and therapeutic relations. But, this defensive position can double up with a more challenging perspective. Users' associations could play a crucial role when it comes to disease mongering because they could change the problem: escaping the ready-made condemnation – the cupidity of the industry, the weakness of scientific 'proofs', the credulity of the public – and questioning the machine itself, that is to say, what makes its parts work together.

If I am considering this possibility, it is because I have had experience of what users' associations were able to do in a slightly different domain, that of illegal drugs. The landscape was defined in that case by the consumers' radical incapacity, almost by definition, to intervene in the question of their consumption. Laws defined them, like minors, as having to be protected against themselves. The alternative proposed to them was either to be condemned as delinquents or to present themselves as sick people in need of care. The presence of those who defined themselves as unrepentant users was crucial. Even if drug users involved in self-help associations were in the minority, the very possibility that they were able to get together to think, learn and fight forcefully changed the perception of many professionals involved in drugs and at the time brought about, if not a change in the law, at least what is sometimes called a 'change in the paradigm', with the formulation of a question which, retroactively, is blindingly obvious: what if laws banning drugs were largely responsible for the catastrophes that are attributed to 'drug taking'?

The most obvious feature of this change is that the problem posed by drugs defined as illicit was 'demoralized'. This has two senses: the healthcare perspective was no longer dominated by the moral principle 'just say no to drugs'; and those going to war for the ideal of a 'world without drugs' were demoralized. Drug use will not be eradicated. The problem posed then concerns the kind of coexistence that can be, or could be, created with these drugs. How does one live with this interesting but formidable power called drugs?

The space opened up for this question has proved fragile. The drug question might have been well and truly 'demoralized' and have thus escaped the empire of psychoanalysis, but it has been captured by other protagonists, weaving new possibilities for medication within a new burgeoning psychopharmaco-cognitivist field of research bearing on addiction and *craving*, progressively including all kinds of 'drugs', licit or illicit.[8] Morality no longer has its hands on the controls but the progress of science does. The fact remains that the event did happen, with its questions, imaginings (what Félix Guattari called the eruption of new collective articulations of enunciation), and this is what I want to bear witness to so as not to ratify failure, so as not to double the reasons which result in its normalization. So it is this event that I thought of when it came to disease mongering, that is to say, to those drugs that, far from being prohibited, are presented as medicines, the fruits of scientific progress in the service of humans.

Let me then propose that condemnation, which brings with it an allusion to the desire to eradicate what is denounced, is a trap. New, licit, industry-produced drugs will not be eradicated, as they sweep through the prescription economy by confronting the doctor with demands from people who are no longer quite 'patients' since they get involved in suggesting their own treatment, even claiming the right to it. They know how to reply point by point to the doctor's objections (they learn it on the Web). We will not return those people, who claim the right to use these drugs, to the role of docile sufferers obeying the knowledge of the therapists who are the only ones capable of defining an authentic illness and prescribing an adequate treatment.

Hands off!

Here, I want to risk a second parallel with another political event: the battle for abortion rights. Here again, women who had an issue with a law, and with a moral imperative in the name of which this law asked them to accept and submit, learned to get mixed up in something that was not their business. Women took in hand that which was supposed to define them. The well-known slogan, 'My womb belongs to me', could certainly be discussed for its simplicity and individualism. But if we understand what started the women moving, it was more like 'My womb does not belong to you', and there any individualist simplification disappears. It is a real 'hands off!' shouted at all those who, in the interests of the state or of morality, want to take charge of women's wombs.

In parallel, we can understand those saying 'We

know, we are the only ones who know what we suffer from' as first signifying, 'Hands off! What we are suffering from doesn't belong to you.' We need help (maybe, maybe not) but this question of help does not entitle anyone to capture us, enlist us in the service of their theory or transform us into cases referred to the statistical groups which authorize your so-called proofs. We don't have to depend on your theoretical conflicts, whether your theories identify us as suffering from a 'real' disorder, objectively defined, or as running away from our unconscious conflicts, or even as victims of a society bent on performance or perfection, or yet again as guilty of escaping our responsibilities. We will not be hostages to your interpretations!

Such a position is not just a way for users' associations to differentiate themselves from patient lobbies; it is also and above all a new, non-denunciatory, way of event making, calling into question the illness-vending machine. This machine needs what it destroys, that is, the authority of medical knowledge, the progress over ignorance associated with the verdict 'This is a real disease.' The machine, like the doctors, needs a 'disease-centred' model, that implies that a drug is justified if, and only if, it responds to what has been authenticated as a 'real disease'.

From this point of view, disease mongering, the creation and selling of diseases, is well named because the claim, 'We are suffering from a *real* disease,' mobilizes patients. It is also the reference to real disease that allows the rhetoric of scientific progress, centred on the discovery of basically objective and biological causes of human suffering. But this is what also nourishes the denunciation addressed to the greed of the industry and the credulity of people: if a disease has to be sold, it is because it *isn't* real!

Is it possible to imagine that, when it is a question of psychotropic drugs at least, the action of users' collectives would result in a passage from a disorder-centred model of the illness to a drug-centred model?[9] What would it mean to break the link joining the use of drugs, especially drugs that intervene in overall cerebral functioning, to the difference between uses legitimated by a real disorder and uses (supposedly illegitimate) aimed at escaping a reality that is too hard, demanding, annoying or frustrating? Can it be envisaged that those who take such drugs do it for their own reasons, without the need for justification in medicine's moralizing language: yes, you can take it because your disorder means that you 'really' need it?

Take the case of Ritalin, which many students, especially medical students, take to 'give them a lift'. What is really so objectionable about not passively accepting that there are differences in the ability to concentrate, when schools and examinations make such differences so decisive that careers, if not lives, depend on them? But there will be objections: that can be dangerous! But the whole question then is: do the criteria used to legitimize taking the drug – the fact that it intervenes in a 'real' disequilibrium in cerebral functioning – contribute to the promotion and cultivation of the attention appropriate to these possible dangers?

The example of former associations of drug users defined as illegal is relevant here. These users were the first to recognize that the consumption of drugs is not insignificant. But it was they themselves, they claimed, who were best qualified to put together the knowledge of these risks and transmit what they would have learned in a relevant way to other users who would then know that it was not disguised propaganda to try to persuade them to give up their consumption.

The idea of a 'drug-centred model' is not therefore that of unbridled consumption. It implies the construction of a user knowledge, knowledge bearing on the consumption of drugs as such, on their evaluation and the type of attention needed when dealing with them.

It is not unheard of to look at things from the perspective of risk culture. Amateur enthusiasts of ultra-light aircraft were successful in organizing a detailed study of each accident and the publication of the conclusions. In the case of so-called 'psychotropic' drugs, this perspective conveys the need to address them as powers that can be quite formidable if one doesn't cultivate the skill and precaution required. A starting point is giving up discourses of appropriation: these drugs, even if synthesized in the laboratory, are not defined by a scientific or medical type of knowledge, which means that no one owns the right to define the meaning of their use in the name of moral-medical-scientific reason. The ultra-light enthusiasts know the risks associated with metamorphosis from 'earthbound' to 'airborne' – lots of habits have to be changed to learn to live in the air. Learning what is demanded by living with a drug means, first and foremost, not defining it as a means justified by its end (in particular responding to a 'real disorder') but rather as a journey to be undertaken with a demanding and potent being, one that can be a vector for metamorphosis or an all-consuming power.

The user culture, in contrast to instrumental, diagnostically justified uses, is a problem of collective interest that needs collective knowledge. We can call it a collective expertise in the old sense where the expertise first designates knowledge coming from experience and is cultivated in its relations with experience. Here, this is the experience of the encounter with powers that are not just associated with drugs but presumably with all efficacious psychotherapies. And this experience has a vital need for its own kind of knowledge that user associations can construct. This knowledge is valuable in itself, but in addition it can make other knowledges recognize that they are all gathered around something – a being? a power? – that belongs to no one, that no one can appropriate or represent.

The pharmaceutical companies are well named. With *pharmakon*, the Greeks effectively named not only drugs but the assemblage of what is characterized by an indeterminate power, that is, a power beneficial or formidable, according to its uses. From this point of view, one can say that it is not only drugs that are *pharmaka* but the whole set of powers put in motion by psychotherapies that are involved in what is never just a 'fixing up', that never satisfy the model of the scientifically or rationally 'ideal' disease – for instance, the infection that antibiotics are able (but for how long?) to cure – but rather a transformation, a metamorphosis. It should be added that referring to treatments as aspiring to being scientific, or rational, is the worst of responses to the *pharmakon* question. Appropriating a *pharmakon*, that is, claiming one can define what it 'really' is, means rendering oneself incapable of what it is asking for, that is, a culture of uses.

In the course of this journey, I have not abandoned what it was that triggered my initial thoughts, the possibility of a medicine that, shall we say, walks on two legs. Having been through the user lobbies test, this possibility has become a bit more precise. Illness has become a spectrum.[10] At one end of the spectrum, the question of 'sales' does not apply because the illness has no need of a lobby in order to get recognition. Patients' expertise and medical knowledge don't have to battle about diagnostic methods and medical treatments, rather they negotiate their respective concerns and priorities. At the other extreme, where in the name of science, psychiatrists, psychoanalysts and pharmacists 'sell' definitions of 'what's wrong' in people's lives, and the treatment that this definition justifies, it is well and truly a case of reclaiming a collective and practical expertise. In the end, it is not so much the question of the 'sale' that matters in this field because in their own way all healers 'sell' the beings whose intervention they are diagnosing, in the sense that, as Tobie Nathan constantly demonstrates, the diagnostic's primary function is to make the illness accessible to the treatment; it is made, strictly speaking, a part of the treatment. The problem only arises when the diagnosis is made 'in the name of science', and the pretension for a 'true definition' (monotheistic, in a way) fabricates converts and missionaries.

The disease-mongering machine's capture and redefinition of ongoing operations have amplified the tensions that inhabit this field. It has begun to look like a caricature of a battlefield, where pretensions, condemnations, lobbies and converted patients clash, some clamouring that psychoanalysis has saved them, others that it was behaviour modification therapy, yet others advocate the latest formula™ ... But the process of caricaturing destroys the dream of a return to the normal, to a respectable model centred around an illness which progress should ultimately identify and treat. That is why user associations are crucially positioned. They are not asked to situate themselves as inheritors of a history of progress but rather as inheritors of what has been excluded in the name of progress, expropriated or disqualified as opinion or superstition. This is what American activists mean when they speak of *reclaiming*: reappropriating not a position of authority ('my womb belongs to me!') but the capacity to escape impotence and resist what has fabricated impotence ('my womb does not belong to you!').

4

Doctors, Healers, Therapists, the Sick, Patients, Subjects, Users . . .

Tobie Nathan

Next in wisdom to that, is this other custom which was established among [the Babylonians] – they bear out the sick into the market-place; for of physicians they make no use. So people come up to the sick man and give advice about his disease, if anyone himself has ever suffered anything like that which the sick man has, or saw any other who had suffered it; and coming near they advise and recommend those means by which they themselves got rid of a like disease or seen some other get rid of it: and to pass by the sick man in silence is not permitted to them, nor until one has asked what disease he has.[1]

Herodotus, *The Histories*, I:197

In parallel, we can understand those saying 'We know, we are the only ones who know what we suffer from' as first signifying, 'Hands off! What we are suffering from doesn't belong to you.' We need help (maybe, maybe not) but this question of help does not entitle anyone to capture us, enlist us in the service of their theory or transform us into cases referred to the statistical groups which authorize your so-called proofs. We don't have to depend on your theoretical conflicts, whether your theories identify us as suffering from a 'real' disorder, objectively defined, or as running away from our unconscious conflicts, or even as victims of a society bent on performance or perfection, or yet again as guilty of escaping our responsibilities. We will not be hostages to your interpretations!

Isabelle Stengers, pp. 151–2, this volume

First, there is the matter of vocabulary. How do we name ourselves and name those we work for? Therapist? Healer? Mercenary? . . . And them? Patients? Clients? Subjects? Or users?

Therapist

In traditional Jewish thought, it is often said that therapeutic practice, the art of healing, is the first level of what is called the 'practical Kabala', which is to say religious knowledge applied to everyday life . . . . The therapist's work is therefore that of the beginner, generally reserved for the apprentices, those not yet capable . . . . Yet it is still work, of course, but it is a kind of introduction to the real movement of thought. Fancy that! I've done it for thirty years! After that time, by any logic, I should have reached the second level . . . . In any case, I don't know what is at the second level, if it exists. The obvious question remains, why does Jewish religious thought give so little importance to therapy, while at university they taught me the opposite – particularly in the case of psychotherapy? At that level, you needed a consecration to access therapy as the highest level of knowledge, an apotheosis, so to speak. First, we were taught the theory and then, at the very end of the course, we were finally authorized to act, to do something. The movement was therefore from ideas to matter, supposedly in a kind of logical progression, abstract to concrete, as if it hadn't been noticed that the hands learn by doing; and that knowledge – and especially practical knowledge – comes more easily through the eyes than the ears. Afterwards, that was something I often noticed with the healers that I had the opportunity to meet in Reunion, Mali, Benin, Burkina Faso and Brazil. For them, the therapeutic act, strictly speaking, is also subaltern work that they are happy to leave to young apprentices. It is these young ones who learn silently, who seek out the ingredients, fabricate the amulets, sew the pieces of leather. The master reserves speech for himself – not speech for the patient but speech that crafts the order of the world. That is how it is! Plato has infiltrated our teaching, making the apprenticeship for therapy so difficult in our domains.

The sick

As a therapist, I have lived through three epochs and three ways of designating the people with whom psychiatry has occupied itself. When I started off, right at the beginning of the seventies, they were still called the 'sick'. I thought – I still think! – that it was a good thing to call them this, no doubt the least of evils, knowing the force of the designations and consequences of the naming that have followed. Calling them 'sick' meant it was recognized that there was a being that accompanied them, a force that was a stranger to them and yet that ruled over their existence: the sickness. And when the presence of such a 'being' is established, humans always gather around to acknowledge it, study it, discover ways to tame it, ways to introduce it into the world of humans. These beings, at least, can be political . . . unlike the 'shrinks'!

Consider what happens when an aetiology of a biological kind is found; each time the sick are liberated, groups are constituted, based on similarity, and solidarity processes are set in motion. Everyone was waiting for the same thing to happen for psycho-pathological disorders. They were expecting to find a simple being, a type already known, like a germ, bacteria or even a gene. Unfortunately, this being remains to be found in most psychiatric pathologies, and Philippe Pignarre is right to reject the pretensions of this psychiatry that is only 'biological' in its distant intentions; psychiatry that can only allow itself a 'petty biology', just as its clinical scales are based in a tiny 'petty psychology'.[2]

I also remember that when people were designated as 'sick', there was a side to it that was quite like a police inquiry: look for the being hiding behind the unusual pallor of a fiancée, for example. This procedure can be found in Freud's letters to Martha – Freud wondering if his fiancée is not affected by an 'illness' transmitted from her forebears. Having the sickness was in a certain manner also being it . . . And it is the same logic in Senegal or Mali where, if there are spirits who have possessed the women of a family one or two generations earlier, there is a good chance that these spirits re-emerge in the following generation. Sicknesses do not just class people; they also draw up genealogies.

Patients

In my own 'psych' context, I often used a second label, 'patients', like everyone else who was in the habit of using it without thinking too much about its semantic implications. The word comes from the Latin *patientum*, from *patior*, 'to suffer'. No doubt it was a way of saying 'what interests me about him is his suffering'. And, of course, the 'shrinks' had something else in mind. They said to themselves that they would end up convincing the 'patient' in question that he or she had 'chosen' to suffer. I met endless numbers of people who had suffered a 'nervous breakdown', migrants for the most part, workers who had had a workplace accident and continued in pain despite treatment, despite X-rays that showed no sign of lesions, despite other scans ... in the end they were sent to a shrink, which made them complain: 'You are sending me to a psychiatrist .... Are you implying that I am doing it on purpose, that I am making it up?'

These are the characters fabricated by the 'psych' disciplines who suffer so much they end up realizing it was 'their desire', they, the 'patients'... There are more subtle traits as well. I hadn't thought about it at the time, but today I can easily recognize it: it is the sacrificial lamb. As anthropologists know, at least since Marcel Mauss, no animal is sacrificed without agreeing to be sacrificed! An animal offered to a divinity is always a volunteer. And how do we know that an animal agrees to die? A chicken, for instance, that someone wants to sacrifice to a save a sick person – how do we know that it will go along with being offered as a sacrifice? Often it is the case that a blade of grass is slipped into its beak, or perhaps a few drops of water. Then, if the chicken eats the blade of grass or swallows the water, then it consents. It agrees to be the sacrificial animal.[3] But where the chicken spits out the blade of grass, or rejects the water, then it is refusing! In that case, it is not killed; it is allowed to carry on its life as a chicken. I must say that it is a rare occurrence. We know this kind of thing in our world as well, marked as it is by Christianity, especially with the story of Jesus sacrificing himself to save mankind, something that gave him the status of 'the lamb of God'. By delivering himself voluntarily to save humanity, Jesus does nothing more than follow the pattern of animals that consent to sacrifice.

So this is what we were looking for with our sufferers when we used the term 'patient' for them. We wanted them to be consenting, submitting to their destinies as lambs of God. I even know how this consent was tested! The equivalent of the blades of grass or drops of water was the requirement to pay. If they agree to pay, the thinking went, that means they consent. They take on the role of sacrificial animal. But in this case the animal is confused with the sacrificed. This is not the person through whom humanity will come to be saved, only his or her own self. For me, this is the only explanation for the strange financial custom surrounding psychoanalytic practice. All one needs to know is that obscure religion for which so many sheep were sacrificed; and, above all, who this god was who so loved feeding off the suffering of people.

In my career as a shrink, I have encountered this sequence in psychoanalytic circles again and again: people who keep on suffering until the moment when they 'become conscious' that they have chosen their suffering, that this was what they 'wanted'. It is even the paradigm of the famous 'case study'[4] genre, as defined by the first psychoanalysts, and subsequently always reinforced. The cure begins with a kind of rebellion – 'Father, Father, why have you abandoned me?' – and ends with an acceptance: 'I accept this suffering that propels me to the avant-garde of humanity, conscious in the heart of an unconscious people, equal in a group of chosen people who have agreed to submit to the "Law".'

Subjects

The third word used to designate patients, one I find the most deceptive but in a way the most optimistic, is that used to assign them the status of 'subjects'. There is no point spending time on the lack of precision of the term 'subject'; philosophers have shown better than I the false character of this designation. I would simply like to note how much this qualification is a sort of optimistic tautology. The sequence goes something like this: 'In my train wreck of a life, the only light, the one which allows me to keep believing I am a human being, is my decision to come to see you [*toi*], to consult with you as my therapist.' It is eally a very optimistic statement. It is this type of procedure that is evoked when people who come for consultation are designated as subjects. How do they show themselves to be subjects? How does the light illuminating their lives turn on? Through a sign, through the fact that they come to see me! That is what a subject is: a person in distress, in a bad way, suffering. But deep in the shadows of this life, a light appears announcing, 'Here is a therapist!'. And suddenly a subject appears!

We have seen how they are called 'sick', 'patients', 'subjects'. . . And today, people have begun using another term. A new word has arrived: 'user'. But we have to be careful; 'users', in the plural, designates people and not just someone who speaks as a 'subject'. Françoise Giroud, for example, talking of her psychoanalysis with Lacan, is not a 'user'. Why not? Because she is talking on her own, she speaks in her own name, for herself; she offers herself as a model. She is a case study incarnate. In no way is she a member of a group. Marie Cardinal, narrating her psychoanalysis, is not a 'user' either. She is a witness, indeed a martyr, but certainly not a 'user'!

On the other hand, when people get together around a question, and this question becomes political . . . the AFSGT[5] might well have been the first to raise awareness on the specificity of the internal life of people suffering from Tourette's syndrome, and they demanded a modernization in treatment. Then UNAFAM[6] buttonholed members of parliament on the subject of pharmaceutical laboratories' involvement in the choice of psychiatric treatments . . . or, more recently, when associations of parents of autistic children, such as Autisme France, demanded specific care facilities that used demonstrably effective techniques . . . we saw a social group appear that was concerned with singular beings, a political group. Here, we can speak of 'users'. This type of group, these users, are in no way solitary heroes or martyrs to one therapy or another, as occurred during a century of case studies or patient autobiographies. It's quite the opposite!

Users

'User collectives', Isabelle Stengers has called them, and I like the word 'user' when it applies to people making use of therapeutic systems. It restricts players in the therapeutic drama to social positions. It designates the therapist as 'service provider' and consequently responsible, that is, active and open to accountability. It encourages the patient to think of her or himself as a consumer, hence paying attention to what is on offer and considering, even comparing, these offers. It refers to the social body in which 'users' circulate, allowing for the identification of interest groups that always form around the distribution of services. For all of these reasons, the simple fact of adopting the word 'user' clears the ground, opens up perspectives, allows comparisons and authorizes people to think on their feet.

It is a good idea to remember here that 'user collectives' have been around for a very long time, probably since ancient times! It is sufficient to remind oneself of the beginning of Euripides' *The Bacchae*. Dionysus was a strange god, both Greek and foreign, insider and outsider, so to speak.[7] He decided to run his cult out of Thebes. He made himself the promoter of the service he expected of humans. We haven't thought enough, it seems to me, about the modernity of this god as principle agent of his own enterprise. As soon as he arrived in town, he took women on a frenzied race up the mountain. Once there, they devoted themselves to 'mysteries' that we learn from the start of the play are the hunt for wild animals, the immolation and laceration of a billy goat and other animals, the eating of their flesh and blood, the consumption of milk, honey and wine, and no doubt activities of a sexual nature. The old men of the town, represented by Cadmus, the founding father, and Tiresias, the blind prophet, decide to take part in the ritual.

Cadmus: Of all the city are we the only ones
who'll dance to honour Bacchus?

Tiresias: Yes, indeed,
for we're the only ones whose minds are clear.
As for the others, well, their thinking's wrong.[8]

But the king, Pentheus, takes fright at the disorder and wants to prohibit the cult of the new god. But this goes badly for him. Hidden high in a pine tree to spy on the Maenads' frolics (the women of Dionysius' cult), he is spotted by his own mother who, seeing him from far off, mistakes him for a lion cub. She falls upon him, tearing off his arm with her own hands, and then his head. The king of Thebes was not only assassinated by his own mother but treated as a sacrificial animal.[9]

Euripides has written a very good description of the sequence whereby user collectives are constituted. Dionysius is the new force suddenly imposing itself in the city. It is unknown, therefore strange. It brings new obligations with it. Established rituals can neither be drawn from it nor applied to it. If it is ignored, then women, as 'Maenads', will roam madly over the mountain. Catastrophes come from not dealing with it according to the rite. Political power wants to deny its existence and prohibit the gatherings that it brings with it. The king, Pentheus, even tries to imprison the god.

We know that the introduction of the cult of Dionysius,[10] which started in Greece in the sixth century, ended up creating congregations (*thiasi*), in each city that periodically organized the rite. We also know that the rite became progressively more stable and that it involved animal sacrifice, the consumption of substances and cult members in trances. In Greece, Dionysians were both a cult and a popular therapy.[11] Such was the success of the cult that it is generally thought that Dionysus succeeded Zeus as king of the gods.

So here is the how user collectives are constituted, according to the sequence that the Bacchae have perfectly sketched out: (1) the appearance of a force, an ambivalent power (Isabelle Stengers is right to speak of the *pharmakon*); (2) the constitution of a collective concerned about the force, whose principal activity is to investigate the ways in which this force can be controlled; (3) the institution of the force, by means of the collective, in the social environment.

The ancient Bacchae were without doubt such a collective. They brought together women (certainly not only women, but mostly women) around a common resource that was the setting of the Dionysian ritual.[12]

User collectives thus imply the existence of a force – let us use the vaguest term for the moment. In this case, it is a god. But the example is fruitful, permitting us to think, progressively, that religions are not in any way a bundle of beliefs but the code of the user collective that has centred itself around the job of knowing this specific force, this *pharmakon* that has suddenly appeared, this new god Dionysius.

Religions are always procedures for knowing about a divinity, for entertaining the obligations that its presence creates, the services it implies and the precautions to be taken to contain the divinity within a definite perimeter.[13] That, by the way, is what teaches us that divinities are always toxic and that religions are kinds of sets of instructions as to how to manipulate a dangerous substance without too much risk.

Elsewhere, in other worlds, we find this same type of gathering of users in ritual possessions, such as *n'dop* in Senegal, *djinnadon* with the Bambara of Mali, the *zars* cult in Ethiopia and the Sudan and the *mlouks* among the Gnawas of Morocco . . .[14] Here, too, everything begins with disorder, with sickness. When people fall sick, and they are diagnosed (or it is postulated) that they are 'accompanied', it is most important to 'make present' the power. What we designate by the word 'possession' is, for the Maghrebian rituals (of Morocco and Algeria), the word *hadra*, 'presence'. This spirit that they want to see express itself might be a spirit they know already, or perhaps they don't yet know it. It is during the rite, the time of identification, that the accumulated knowledge of the users' collective are indispensable because the spirits – I mean the *djinns, zars* and *mlouks* – behave in specific ways. They like living with spirits, just like humans live with humans.

Now, user collectives, like the Bacchae of old or the *n'dopkat*[15] of Senegal today, take this need that spirits have into account. They suggest moments to humans during which they can live the life of spirits (*rabs, djinns* or *mlouks*). For these moments they become spirits, or at least lend their bodies, speech and thoughts to the spirits, at least for the time of the ritual. And since what is involved is reconstituting an ambiance, a tonality of life conducive to the ecology of spirits, this can only be done by bringing together people who themselves have experienced spirits. In this way, new spirits recognize the presence of other spirits that have lived with humans for longer, feel welcomed in a highly charged world of spirits and will in turn manifest themselves. That is why those rituals that we are in the habit of designating as 'possession', need to have, in order to proceed, the presence of people used to possession, that is, the former possessed.

So these really are user associations, but associations made necessary because of a kind of requirement on the part of forces that these associations have given themselves the task of exploring. The former possessed should regularly grant a place to spirits with whom they once concluded some alliance, and the new possessed need the possession of former ones to attract, then train, their spirit that is still shy, still wild, hesitating to set foot in the world of humans. Except that in such groups they don't talk about 'users' – I mean that the conceptualization is not centred on humans. The ritual mechanism is described from the perspective of the beings who invest the users, the beings called *djinns, rabs, zars, mlouks* or *haoukas*.[16]

And with the simple fact of the ritual taking its course, the humans are constituted, under the pressure of functional necessity, as a user collective – and this is because the force in play, the *pharmakon*, is precisely another substance, unknown to humans.

We are certainly dealing with 'users' and not 'the sick' because it is very difficult to see how such a gathering differs from any religious grouping that gets together to honour a divinity or from a family or religious assembly that has decided to commemorate some date. In addition, the treatment of the *pharmakon*, the ambivalent substance of the god, the being that could turn out murderous, concerns the whole group. The users' collective consequently feels not just concerned about the *pharmakon* but also involved in a task that is useful for the larger group.

Pharmaka

Now we need to know what we might consider, today, as the equivalent of these forces. What powers are comparable to the irruption of a god in an ancient city, powers that demand by their *presence* the formation of a group of concerned people that gives itself the task of exploring the ecology of this force?

In the first instance, it is a matter of sicknesses, like in antiquity, and on this point the note from Herodotus on the Babylonians seems to me prophetic. Herodotus knew very well, through his description of a no doubt imaginary Babylon, that sicknesses only exist to the extent that they constitute a necessarily unified group of 'sick people'.

Now, once these sicknesses are constituted as objects of investigation by collectives, they are deployed and recruit new adepts. There are several explanations for this phenomenon, at first seemingly paradoxical. These illnesses are only successful to the extent that they find these collectives. Looking just at psychiatric syndromes, I remember that during my studies we learnt long lists of categories of 'psychotic disorders', from the 'delusional episode' to 'chronic hallucinatory psychosis' – there were more than a hundred. These categories, which at the time reminded me of entomological classifications, evaporated because, being just for professionals, they couldn't be at the origin of the constitution of collectives. Some fruitful categories, although polemical ones, remained at the centre of public debate. Depression, which scarcely existed at the beginning of the 1970s, at least in textbooks which spoke more about melancholy and neurasthenia, has seen a spectacular rise to the point where Philippe Pignarre can quite rightly speak of a real epidemic.[17] But depression has not been able to pull interest groups together around it. It has interested pharmaceutical laboratories and produced an abundance of wordy dissertations by sociologists with a need for a research topic. Bipolar disorders, on the other hand, are on the way to bringing together quantities of user collectives – users of the illness but also of the medicines, the different categories of mood regulators, new surgical propositions, reflections on the personalities engendered by the unique amalgams of people and products. Autism is a paradigmatic example of how successful an illness can be on the basis of its capacity to bring groups of concerned people together – in this case, associations of parents of autistic children. In the 1970s, the statistics showed one case in ten thousand; in the 1990s one case in a thousand. In addition, the mobilization of parents' associations has succeeded in getting the HAS (*Haute Autorité de Santé*) to counsel against the application of psychoanalytically inspired methods in the treatment of autism. This event is actually very important historically since the decision went against the advice of the majority of professionals and the political and media pressure groups. It is the turning of a corner and allows us to predict some profound restructuring of the field of mental health in France. The same applies to schizophrenia and those psychotic disorders that have not succeeded in bringing together interested persons, despite the remarkable breakthrough of Deleuze and Guattari's books on schizanalysis in the 1970s that attributed to schizophrenia a capacity for the radical deconstruction of the system of modern thought.

However, it is via a more secondary route, as far as professionals are concerned, that troubles of a psychotic nature have shown themselves capable of bringing together interested people: voices. What has emerged, first in Switzerland, then in France and Belgium, is the Hearing Voices Movement that re-evaluates the idea of living with voices which suggests that certain benefits can be drawn from it and that it is not always desirable that this singular capacity should be quashed by particularly abrasive neuroleptics. In short, for psychopathology, sicknesses can gather forces to themselves, just as in the *pharmaka*, and turn out to be likely attractors of user collectives. But the analysis of recent movements shows us that it is not the sicknesses, as defined by the 'experts', but neighbouring categories, extending and not always overlapping with those of the experts and capturing older, 'traditional' thought long ago predicted to have died out. Those who are hearing voices provide a striking example, linking up with millennial preoccupations with non-human perceptions. But the collectives gathered around autistic children are nevertheless passionately interested, and in a practical way, with the irreducible singularity of these children, so different that they need special education in order to direct them towards a life with the broader human society.

So I agree with Isabelle Stengers's definition of the *pharmaka*, constructed on a drug model, but, besides medicines, I would add two other types of beings: *gods* and *sicknesses*. I consider these beings to be just as dynamic as drugs, perhaps just as active as them in the construction of persons, especially their identity.

So what becomes of the therapists' role in a world that ceded, or will cede, its place entirely to user collectives? They will remain experts of a kind, no doubt, but certainly not the only ones. Collectives constitute another kind of expertise that will have questions to ask of the therapists. They will no longer be able to refer back just to their own professional group. They will have to share their points of view, bring them to the table with those of the collectives. But the most sensitive changes will take place with their techniques.

In one way or another, they will have to integrate the collectives into the very space of the consultation. We have already experimented with this way of working for about thirty years, ever since the first ethnopsychiatric sessions where we always brought experts from the patients' worlds into the space. Henceforth, this way of working will not just be for 'ethno-shrinks', it will necessarily be that of all shrinks, who will have to introduce contradiction in the very place where the 'word of the master' used to reign.[18]

Notes

Chapter 1   Towards a Scientific Psychopathology

1 It is astonishing to note how many of psychoanalysis's so-called 'sexual interpretations' are unknown
only to the solitary sufferer. In the case of Dora,
for example (Sigmund Freud, *Dora: An Analysis of
a Case of Hysteria*, New York: Touchstone, 1963),
Herr K. knows that she wants it; it's just that she
doesn't want it enough. As for Frau K., she would
naturally have to be Dora's father's mistress. On
the basis of these typical descriptions of these
kinds of patients, one could say that 'everyone
knows' that they are feeling sexual desire except
them . . . .

2 *Babalawo*, in Yoruba, 'father of the secret'. Cf.
Bernard Maupoil, *La Géomancie à l'ancienne Côte
des esclaves*, Paris: Institut d'ethnologie, 1988;
Tobie Nathan and Lucien Hounkpatin, 'Oro Lè.

La puissance de la parole ... en psychanalyse et dans les systèmes thérapeutiques yorubas', *Revue française de psychanalyse* 57(3), 1993: 787–805.

3 Tobie Nathan, 'Frère et soeur: l'amour sorcier (une illustration clinique)', in *Autrement* 112, 1990: 116–24.

4 Lucien Lévy-Bruhl, *How Natives Think*, trans. Lilian A. Clare, 1926 (1910), London: George Allen & Unwin.

5 Three very different props for mounting a divinatory procedure very common throughout Africa. It establishes a correspondence between the 256 combinations of a system of binary oppositions and 256 word-phrases in a secret language. See Robert Jaulin, *La Géomancie. Analyse formelle*, Paris: Mouton, 1967.

6 This is a clinical example of a 'witch child' from Zaire that I gave in Marie-Rose Moro and Tobie Nathan, 'Ethnopsychiatrie de l'enfant', in Serge Lebovici, René Diatkine and Michel Soulé (eds), *Nouveau Traité de psychiatrie de l'enfant et de l'adolescent*, new edn, 4 vols, Paris: Presses Universitaires de France, 1999.

7 African cultural systems seem to posit that twins, especially heterosexual ones, are simply imperfect 'cultural beings'. They are thought capable of recognizing each other, traversing cultural barriers of language or particular customs, perhaps even having secret meetings. In sum, they are always suspected of belonging to the 'sorcerer' group.

8 No doubt it is the 'law', which certain psychiatrists call castration, to which they proudly submit themselves, as well as their patients.

Notes to pages 16–19

9 Isabelle Stengers, *The Invention of Modern Science*, trans. Daniel W. Smith, Minneapolis: Minnesota University Press, 2000; Isabelle Stengers, *La volonté de faire science*, Paris: Les Empêcheurs de penser en rond, 1992. One passage from among so many possible ones in the latter, very stimulating book reads:

> I said that speaking of science was a matter of engagement. So it is on this point that I situate my engagement. I maintain that – contrary to the assumption of epistemologists who hold that an objective statement is a right to which any rational scientist may pretend – the possibility for any science to accede to the enviable 'hard science' status is of the order of an event, which happens, but neither by decree nor by merit. In doing this, I deliberately put in question sciences that have sought to 'merit' this status by mangling their object (behavioural psychology), or by forgetting it (econometrics), or rather I refer their history to another register, where the dominant interests would be academic, economic and political. (p. 29)

In the long list of possible 'pseudosciences', 'knowledge that mimics "hard science"', one could add psychopathology, which simply forgot to construct its object.

10 Sybille de Pury-Toumi, Tobie Nathan, Lucien Hounkpatin, Hamid Salmi, Jean Zugbédé, Constant Houssou, Gilberte Dorival, Souren Guioumichian and Nathalie Zajde, 'Traduire en folie. Discussion

off

181

linguistique', *Nouvelle revue d'ethnopsychiatrie* 25–6, 1994: 13–46.

11 Spoken in Benin and Togo.

12 Mostly spoken in Senegal.

13 Spoken in Congo and Zaire.

14 Spoken in the Central African Republic.

15 We hardly need recall that, in a clinical situation, the therapist's work consists of constructed representations, never research on a supposed objective reality or 'structure'. This is why psychopathology can only be a science of therapists' techniques.

16 This is a verbatim transcription of the dialogue. In such a group, at least one of the therapists takes word-for-word notes and sometimes they are videotaped. See Tobie Nathan, *Fier d'avoir ni pays ni amis, quelle sottise c'était. Principes d'ethnopsychanalyse*, Grenoble: La Pensée sauvage, 1993.

17 Traditional medicines: powdered bark from trees, leaves to make tea or baths, talismans . . . .

18 Scarifications.

19 'Eat': an attack through sorcery . . . devouring the vital substance.

20 Animal sacrifice is a crucial element in the therapeutic apparatuses of African systems. Generally, it is a matter of capturing 'the breath of life' from flowing blood, of acquiring an excess of vitality. Sometimes the blood is not made to flow. If the animal is suffocated, it is the opposite; the breath is enclosed. In a case like that, the animal is substituted for a person in danger. That kind of ritual act comes about to underscore the idea that a human life is at stake. Cf. Henri Hubert and Marcel Mauss, *Sacrifice: Its Nature and Function*, trans. W. D. Halls, Chicago:

University of Chicago Press, 1964 (1899). And also the journal, *Systèmes de pensée en Afrique noire* 2–4, 'Le Sacrifice', 1976–1980.

21 Ethnopsychiatric consultations have progressively refined their methodologies. They began in 1979–1987 in the psychopathology unit at Avicenne hospital, at the time under the direction of Professor Serge Lebovici; they then continued (from 1987 to 1992) at Protection for Mothers and Infants at Seine-Saint-Denis. From 1993, they took place at the heart of Paris VIII University at the Centre Georges Devereux, a university centre giving psychological help to migrant families. The centre is now located in Paris (see ethnopsychiatrie.net). The technical procedures are discussed in Tobie Nathan, *Fier d'avoir ni pays ni amis*, op. cit., and some theoretical perspectives and the results of clinical discussions in Tobie Nathan, *L'Influence qui guérit*, Paris: Odile Jacob, 1994, and a great number of clinical cases in different numbers of the *Nouvelle revue d'ethnopsychiatrie*.

22 I feel it is just as difficult to designate industrialized western societies, based on an observable external description. They have to be qualified by the thoughts they cultivate about themselves. They are societies traversed by a line of force that is rooted in Greco-Roman antiquity – a world confiscated from its first owners, the gods. On the traditional opposition between those who think and those who believe, one can easily refer to all the great thinkers of the nineteenth century: Bachofen, Tylor, Morgan, Marx, Engels and, at the head of them all for the subject we are concerned with here, Sigmund Freud

(*Totem and Taboo*, 1912; *Introductory Lectures on Psycho-Analysis*, 1915–17, etc.). Bruno Latour made a very good and entertaining presentation of this classical opposition in a lecture he gave in October 1994 at the first Colloque internationale d'ethnopsychiatrie.

23 Cf. Freud and his famous comparison of the child singing to himself in the night. He did nothing to change the world, but he was less frightened of it.

24 *Purity and Danger: An Analysis of the Concepts of Pollution and Taboo*, London: Routledge, 1966, p. 59.

25 Marcel Mauss and Henri Hubert, *A General Theory of Magic*, trans. R. Brain, London: Routledge, 1972.

26 Claude Lévi-Strauss, *The Savage Mind*, Chicago: University of Chicago Press, 1968.

27 Bruno Latour, *We Have Never Been Modern*, Cambridge, MA: Harvard University Press, 1991.

28 Certainly a part of Claude Lévi-Strauss's thought; see *Structural Anthropology*, trans. Claire Jacobson and Brooke Grundfest Schoepf, New York: Doubleday Anchor, 1967.

29 An effect they have nothing much to say about! See Philippe Pignarre's article, 'L'effet placebo n'existe pas', http://www.recalcitrance.com/placebo.htm

30 Certainly a part of the thinking of my respected former teacher, Georges Devereux, *Essais d'ethnopsychiatrie générale*, Paris: Gallimard, 1970.

31 Most classical psychoanalysts, beginning with Freud and throughout his oeuvre!

32 A truly amazing discovery! Going through such a developmental procedure to come up with such an

obvious fact is certainly revealing about the poverty of our psychopathological thought.

33 A language group from Senegal. It would not be much use to cite texts explicitly, as it is so easy to refute this type of frequently appearing argument.

34 Cf. Freud's technical remarks according to which, if the symptom is a thread, and if it is pulled right out, with enough patience one can eventually elucidate the whole person.

35 One example is the case of a Fula child from Mali. She was aged two when doctors diagnosed her with a severe genetic illness for which an extremely complex therapeutic procedure was envisaged, requiring very long hospitalization. It quickly emerged that the issue was the following: balancing the cost of keeping her at home, maybe creating a dead Fula or entrusting her to the hospital for several years, being certain that she would end up being a "living-surviving body" 'enclosing an errant soul . . .'. This case also shows once more that psychopathology should develop as soon as possible a coherent theory of its relationship with cultural entities because medically related disciplines (in this case genetics) always have a dire need of it.

36 Cf. for example, Jacqueline Andoche's description of healing in Reunion, written with as much precision as sympathy: 'Une désenvoûteuse réunionnaise: Mamie Louisa', *Nouvelle revue d'ethnopsychiatrie* 24, 1994: 19–44.

37 I am obliged to pay homage to Marcel Mauss. From the beginning of the twentieth century, he set up the model for what should have been a real psychopathology manual. Its chapters, as Mauss

had indicated in article after article, might have been entitled 'Prayer', 'Sacrifice', 'Magic', 'Gift' . . . .

38 For an analytical sketch of this myth, see Tobie Nathan, *Psychanalyse païenne. Essais ethnopsychanalytiques*, Paris: Dunod, 1988.

39 Louis-Vincent Thomas, *La Mort africaine: idéologie funéraire en Afrique noire*, Paris: Payot, 1973.

40 Sory Camara, *Paroles très anciennes*, Grenoble: La Pensée sauvage, 1982.

41 So was the couple's sterility treated by the white doctor's hormone remedy or by the divination ritual? Clearly, one has to have numerous 'case studies' before arriving at a roughly global picture of the facts, rather than continuing to drone on about 'slippages': is the diviner encroaching on the doctor's territory or vice versa?

42 In psychopathology, this always relates to miming. Psychological tests mime the complexity of the diagnostic equipment of biological medicine. The methodology is actually simplistic. Psychotropic drugs only mime the complexity of the active ingredients in the other biological medicines. All of the 'new' transnosographic chemicals appearing on the market only emphasize the fact that, in psychopathology, we are a long way from having medicines that attack clearly defined targets in the organism. Cf. Edouard Zarifian, *Des paradis pleine la tête*, Paris: Odile Jacob, 1994.

43 There is a vast literature on rites of possession. Yet, for the most part, it tends to be descriptive, that is to say, it takes explicit statements to be thought, and not, as I recommend, as *technical tools*. A useful reference on this topic is András Zempleni,

'Possession and Sacrifice', in Michel Cartry, *Dans la peau d'animal*, Paris, École des Hautes Études en Sciences Sociales, 1967. It is a text that tries to discuss possession rituals from a technical point of view.

44 Of course, doctors study a bit of medicine in the course of their studies (and a lot more later on with sick people!), but above all they learn to 'do medicine'. They are slowly inoculated with the medical *ethos*.

45 See Lévi-Strauss's famous text, 'The Sorcerer and his Magic' in *Structural Anthropology*, op. cit., in which he relates how the famous Kwakiutl shaman, Quesalid, found himself designated as a shaman by the fact that a sick person dreamed of him as the one who would cure him . . . and that is just one of many possible examples!

46 All of the descriptions in so-called 'medical anthropology' that consider divinations to be 'proto-diagnoses' carried out with an intuitive method only show one thing: their total incomprehension of the divinatory process.

47 No doubt because it represents the accumulation of particles containing the fragments of the dead that have built up since the first ancestor. By interrogating sand or 'earth', no doubt all one is doing is calling on particles of ancestors.

48 All these expressions can be found in Africa in this form.

49 Expression found in Antilles, and also in the French countryside. It is used to designate therapists that used to be called 'pansers of the secret' (*panseurs du secret*) – a curious conjunction of names . . . .

50 Of course, contrary tendencies exist. Yoruba

balalawos have an annual international congress bringing together diviners from Togo, Benin, Nigeria and even Haiti and Brazil. But this is, mostly, it seems to me, as masters of rituals to divinities. As far as their therapeutic 'tricks' go, I would be very surprised if they exchanged them with a view to sharing them in the community.

51 There is an idea right through Africa (particularly concentrated in Central Africa: Zaire, Congo, Angola, Cameroons) according to which sorcerers would have a tendency to recruit new members and to meet at night in sorcerers' societies. Certain societies even create secret societies for hunting sorcerers.

52 Pathways evoked by statements like: 'Out of ten healers, there are nine charlatans', or again, 'You have to look for a long time to find the one who suits you . . . .' Apropos, those interested in non-western therapies should begin with the notion of *therapeutic course* rather than *therapeutic apparatus*.

53 This case, under the name Yehia, is briefly presented in Tobie Nathan and Marie-Rose Moro, 'Le bébé migratoire,' in Serge Lebovici (ed.), *Psychopathologie de bébé*, Paris: PUF-INSERM, 1989, pp. 683–748.

54 As far as I know, Evans-Pritchard is the first to have attempted a conceptualization of this world of sorcery (E. E. Evans-Pritchard, *Witchcraft, Oracles and Magic Among the Azande*, Oxford: Oxford University Press, 1937). He relates that the autopsy of the corpse of a sorcerer revealed the presence in his stomach of a ball of hair with teeth (teratoma?). Yet, while the sorcerer is alive, the autopsy of a sacrificed chicken is sufficient. A clinical case of a young woman from Antilles, exactly in accordance

with the Azande model, is in Rébecca Duvillié, 'Approche ethnopsychiatrique d'enfants migrants en milieu scolaire', *Nouvelle revue d'ethnopsychiatrie* 28, 1995.

55 Thierry Baranger and Martine de Maximy have made a comparative study between the system of 'therapeutic trial' and that of the 'legal trial', highlighting the juridical thought in the one and the therapeutic virtues in the other: 'L'enfant sorcier entre ses deux juges', DEA thesis in psychopathology, University of Paris VIII.

56 Story collected by Geneviève N'Kossou during fieldwork. We have a great number of sorcerers' confessions. The most striking, also on tape, is of a young eight-year-old boy from Zaire. This child's case (I called him Mélampous) is the subject of a quick presentation in Marie-Rose Moro and Tobie Nathan, 'Ethnopsychiatrie de l'enfant', op. cit.

57 Clinical case set out in Tobie Nathan, *Fier d'avoir ni pays ni amis . . .*, op. cit.

58 I really cannot understand the emphasis that western psychopathologists put on the notion of incest. For example, in such systems of sorcery, incest is not a concept; often there is not even a specific word for it. It is just the sign of a voluntary transgression of a taboo, a sign of belonging to a group of sorcerers. If a man commits incest with his daughter, it is actually a way of sponsoring his way into a sorcerers' circle . . . This should not surprise us, since a somewhat comparable idea is found in the foundation myth for the Greek city of Mycenae, that of the Atrides.

59 A case is described well in Suzanne Lallemand, *La Mangeuse d'âmes*, Paris: L'Harmattan, 1988.

60 I find ridiculous those western family therapists who beg for the presence of a grandfather or uncle at their sessions. In central Africa, anyone who refuses to turn up to one of these family meetings about sorcery accusation more or less condemns themselves.

61 The first two sessions in this case were the object of a detailed analysis in Tobie Nathan, *L'Influence qui guérit,* op. cit.

62 Muslim healer.

63 Sometimes the totality of the elements of the *continuum* can be gathered up for an interpretation on the subject of the dead. In order to maintain coherence, I will not here go into the use of the dead in African therapies. A summary can be found in Tobie Nathan, *L'Influence qui guérit,* op. cit.

64 I am using this expression in the same sense as one speaks of 'active ingredients' in pharmacology.

65 The western distinction between 'prevention' and 'therapy' should be reconsidered in the light of processes going on in these worlds. In fact, the premise of the existence of non-humans constitutes both a 'prevention' and the basic organization (the matrix) of care settings that are turned to when needed.

66 Except of course those who want to mount a sorcery attack on someone.

67 There is one anthropological approach that implies that these objects can be considered 'symbols'. I can't understand the thrust of this at all since the principal characteristic of these objects is that they are real. They can be seen, or you can even have them made. Every anthropologist doing fieldwork has been able to touch them, even quite often take possession of

them herself. All you have to do is know the real inhabitants of the Parisian suburbs quite well to be convinced that there are quite as many such objects in the region. In addition, whoever is using this type of object attributes to it the same kind of effectiveness that we give to our pills. So, when they speak of symbols, what is their speaking position? Of course, Jeanne Favret-Saada, during her work in the bocage (Normandy), couldn't see them and, considering them useless from the point of view of her theory, concluded that they were potential cases at the top of a structural diagram (preferably a triangle . . .). But we are dealing here with contemporary France which, more than any other country in the world, is awash with the propaganda of the intelligentsia coming from the demi-monde of television and cinema; journalists, popularizers, etc. It is clear – and here she will no doubt agree with me – that in the African systems spell-objects are just as widespread as the techniques and the counter-spell-objects.

68 In this respect, they resemble industrial objects like tape-recorders or laptops, but also a meal. For a first attempt to describe these objects, see Tobie Nathan, *L'Influence qui guérit*, op. cit.

69 Since we are talking about technical principles, many other mechanisms to constitute envelopes might exist or could be invented.

70 There is an introduction to this type of philosophy in Tobie Nathan and Lucien Hounkpatin, 'Oro Lè. La puissance de la parole . . .', op. cit.

71 A Soninke case is described in Tobie Nathan, *Le Sperme du diable*, Paris: Presses Universitaires de France, 1988.

72 For a discussion of this type of language, see Sybille de Pury-Toumi, Tobie Nathan, Lucien Hou003Hounkpatin, Hamid Salmi, Jean Zugbédé, Constant Houssou, Gilberte Dorival, Souren Guioumichian and Nathalie Zajde, 'Traduire en folie . . .', op. cit.

73 The same kind of idea is found in *Le Grand et le Petit Albert*, a strange sorcery manual of uncertain origin but which has become a common guidebook for thousands of francophone healers from the Antilles, Reunion, Haiti and even parts of Africa.

74 See the discussion of this notion in Tobie Nathan and Lucien Hounkpatin, 'Oro Lè. La puissance de la parole . . .', op. cit.

75 For 'good' or 'ill'; for *protection* or for an *attack through sorcery*.

76 I have deliberately excluded from my discussion all references to the use of plants. These are procedures that can be too easily 'assimilated' to scientific thought. I will have to come back to this later in order to re-institute the full complexity of these mechanisms.

77 As is well known, this procedure is systematically utilized in 'Turkish baths' (*hammams*).

78 The placenta is very important in all object-using African therapies. It is never a matter of creating a 'double' of the subject to be acted upon at a distance (as is usually naively understood) but of fabricating a real 'placenta' from which the person's life or death will then flow. For a discussion of such objects among the Malinkés, see Yossouf Cissé, 'Les nains et l'origine des *boli* de chasse chez les Malinké', *Système de pensée en Afrique noire* 8, 1985: 13–24.

79 I repeat: it is as much a matter of protections as 'spell-objects'.

80 According to the western phraseology that was naturally used to clothe his therapeutic journey: 'perverted structure'; 'infantile character'; 'fixated at the sado-oral stage', etc.

81 For all the actors, without exception: doctors, paramedics, pseudo-doctors and also masters of the secret, masters of the hole, pansers of the secret, just like priests, gurus, shamans and other sorcerers . . . .

Chapter 2   The Doctor and the Charlatan

1 Léon Chertok and Isabelle Stengers, *A Critique of Psychoanalytical Reason: Hypnosis as a Scientific Problem from Lavoisier to Lacan,* trans. Martha Noel Evans, Stanford: Stanford University Press, 1992.

2 Philippe Pignarre, *Le Grand Secret de l'industrie pharmaceutique,* Paris: La Découverte, 1995.

3 Jacques Derrida, 'Plato's Pharmacy', in *Dissemination,* trans. Barbara Johnson, London: Athlone, 1981.

4 Isabelle Stengers, *The Invention of Modern Science,* trans. Daniel W. Smith, Minneapolis: Minnesota University Press, 2000 (1993).

5 Bruno Latour, *Science in Action,* Cambridge, MA: Harvard University Press, 1987.

6 See Léon Chertok and Isabelle Stengers, *A Critique of Psychoanalytical Reason,* op. cit.

7 Readers will recall that Milgram set up an experiment where his subjects, thinking they were taking part in an experiment about memory, found themselves

called upon to punish, via increasingly powerful electric shocks, the errors of memory committed by their human 'guinea pigs' (in reality, Milgram's accomplices). The majority obeyed orders and continued to administer the shocks, while the 'victims' screamed and begged for the torture to stop.

8 As in Molière's famous play *Le Médecin malgré lui*, where Monsieur Jourdain discovers he has been speaking prose all his life without even knowing it (trans.).

9 For psychoanalysis, which is first inscribed within a field of 'experimental' rationality, only to later define itself on the basis of an ethics that opposes the 'subject' to the 'object of knowledge', see also Léon Chertok and Isabelle Stengers, *A Critique of Psychoanalytical Reason*, op. cit.; Isabelle Stengers, *La Volonté de faire science*, Paris: Les Empêcheurs de penser en rond, 1992; and Isabelle Stengers, 'Les déceptions du pouvoir', in Daniel Bougnoux, *La Suggestion: Hypnose, Influence*, Paris: Les Empêcheurs de penser en rond, 1991, pp. 215–31.

10 Tobie Nathan, *L'Influence qui guérit*, op. cit.

11 See Isabelle Stengers, 'L'expert et le politique', in Francis Caballero (ed.), *Drogues et Droits de l'Homme*, Paris: Les Empêcheurs de penser en rond, 1992.

Chapter 3   Users: Lobbies or Political Creativity?

1 See Isabelle Stengers and Olivier Ralet, *Drogues, le défi hollandais*, Paris: Les Empêcheurs de penser en rond/La Découverte, 1991.

2 http://collections.plos.org/disease-mongering

3 Lynn Payer, *Disease-mongers: How Doctors, Drug*

Companies, and Insurers are Making You Feel Sick,
New York: Wiley & Sons, 1992.

4 It will be noted that other 'sales' procedures move through distinct strategies in the field of so-called purely somatic diseases, for instance the promotion and sale of treatment for the 'healthy sick', those who present an excess (or lack) of X that is said to 'favour' the development of sickness Y. Here there are huge businesses directed at the public that inform everyone that they need to monitor their degree of . . . .

5 See, in particular, Félix Guattari, *Chaosmosis: An Ethico-Aesthetic Paradigm*, trans. Paul Bains and Julian Pefanis, Bloomington & Indianapolis: Indiana University Press, 1995 (1992), as well as Gilles Deleuze and Félix Guattari, *A Thousand Plateaus*, trans. Brian Massumi, Minnesota, MN: University of Minnesota Press, 1993 (1980).

6 Philippe Pignarre, *Comment la dépression est devenue une épidémie*, Paris: La Découverte, 2012.

7 Children and Adults with Attention Deficit Disorder.

8 See Josep Rafanell I Orra, *En finir avec le capitalisme therapeutique. Soin, politique et communauté*, Paris: Les Empêcheurs de penser en rond/La Découverte, 2011.

9 On this subject, see the very innovative article by Joanna Moncrieff and David Cohen, 'Do Antidepressants Cure or Create Abnormal Brain States?' *PloS Medicine* 3, (July) 2006, plosmedicine. org.

10 A spectrum the medical uniformity of which only user associations can break by exploring its distinct components according to the type of role that they,

as users, can claim and the type of knowledge that they will make themselves capable of constructing. What of the case of Huntington's disease that has undoubtedly the power to impose its own recognition and can be predicted by genetic tests while doctors are powerless in the face of it? As I check over this text, an association called Dingdingdong has started to formulate this question.

Chapter 4   Doctors, Healers, Therapists, the Sick, Patients, Subjects, Users . . .

1 According to the historians of antiquity, there were plenty of doctors in Babylon. Perhaps this custom was only to do with the villages and not the city. Or was Herodotus attracted by this idea of democratic distribution of popular medicine, as he sees in the Babylonians the very model of solidarity . . .?

2 See Philippe Pignarre, *Comment la depression . . .*, op. cit.

3 See Tobie Nathan and Lucien Hounkpatin, *La Parole de la forêt initiale*, Paris: Odile Jacob, 1996.

4 Mikkel Borch-Jacobsen, *Les Patients de Freud*, Paris: Editions des sciences humaines, 2011. He shows in a decisive fashion how most of the clinical cases that Freud reported on were 'fakes', as they say in journalism. We already know this, after the incontrovertible summation by the same author, with Sonu Shamdasani (Mikkel Borch-Jacobsen and Sonu Shamdasani, *The Freud Files: An Inquiry into the History of Psychoanalysis*, Cambridge: Cambridge University Press, 2012), but this time we can't avoid the truth any longer. We also know, perhaps without having considered all the

consequences, that most of the patients taken up by Freud became even sicker, sometimes significantly so. Certain psychoanalytic cures, we learn in this book, turned into tragic affairs. The case study, the literary genre launched by Freud, should be done away with and replaced with user collectives taking charge of the clinical information.

5 French association for people suffering from Tourette's syndrome.

6 National union of families or friends of people suffering with or handicapped by psychic disorders. The UNAFAM site has a list of the various associations for ill people or their families.

7 Marcel Détienne, *Dionysos mis à mort*, Paris: Gallimard, 1977; Maria Daraki, *Dionysos et la déesse Terre*, Paris: Flammarion, 1994.

8 Euripides, *The Bacchae*, Scene 1.

9 Note that the specificity of the Dionysian animal sacrifice is precisely that of a wild animal, that is, non-consenting. It is for this reason that it has to be killed in a savage fashion, lacerated and devoured like prey. Pentheus, as the recalcitrant, is the paradigmatic wild animal in the cult.

10 It was in fact a reintroduction, since the cult of Dionysus, although few in number at the time, had existed since the second millennium BC (Maria Daraki, *Dionysos et la déesse Terre*, op. cit.).

11 E. R. Dodds, *The Greeks and the Irrational*, Berkeley: University of California Press, 1951; *Pagan and Christian in an Age of Anxiety: Some Aspects of Religious Experience from Marcus Aurelius to Constantine*, Cambridge: Cambridge University Press, 1965.

12 See Michel Bourlet, 'L'orgie sur la montagne', *Nouvelle revue d'ethnopsychiatrie* 1, 1983: 9–44, http://www.ethnopsychiatrie.net/actu/bourlet.htm. This is one of the best texts ever written on possession rituals.

13 This, by the way, is the etymology of the word 'sacred' in Latin, Hebrew and Arabic.

14 For *n'döp*, Andràs Zempleni's works are essential: for example: 'La dimension thérapeutiques du culte des rab: n'döp, Tuuru, et Samp. Rites de possession chez les Lebous et les Wolofs du Sénégal', *Psychopathologie africaine* II(3), 1966; see also his thesis 'L'interprétation et la thérapie traditionnelle du désordre mental chez les Wolofs et les Lébous (Sénégal)'. For the *djinnadon* in Mali, see Jean-Marie Gibbal's book, *Guérisseurs et magiciens du Sahel*, Paris: Métailié, 1991. For the *zars* cult in Ethiopia, see the early text by Michel Leiris, *La Possession et ses aspects théâtraux chez les Éthiopiens de Gondar*, Paris: Plon, 1958. For the study of *zars* in the Sudan, see Sadok Abdessalam's thesis, 'Le Voleur et le Visiteur. Analyse de deux systèmes thérapeutiques (le djinn et le zar) au Soudan, dans la région de Gézira, thèse pour le doctorat d'ethnologie', University of Paris VII, 1993. For the Gnawas of Morocco, see Abdelafid Chlyeh's thesis, 'La Thérapie syncrétique des Gnaoua marocains', doctoral thesis, University of Paris VII, 1995.

15 *N'dopkat*: members of the brotherhood of *n'dop*. The film made by Andràs Zempleni and Henri Collomb, *Le N'döp*, is available online.

16 On the topic of *haoukas*, see the ground-breaking

film that Jean Rouch made in Ghana on the *haouka* brotherhood, *Les Maîtres-fous.*

17 Philippe Pignarre, *Comment la depression* . . ., op. cit.

18 See Tobie Nathan and Nathalie Zajde, *Psychothérapie démocratique*, Paris: Odile Jacob, 2012.